S0-BWX-191

Ambient Energy and Building Design

Ambient Energy and Building Design

Edited by
J E Randell *BSc PhD MCIBS,*
Dept of Civil Engineering, University of Salford

The Construction Press
Lancaster, London and New York

This book is based on the conference of the same title, organised in April 1977 by the Construction Industry Conference Centre Ltd in conjunction with the Royal Institution of Chartered Surveyors, the Chartered Institution of Building Services, the International Solar Energy Society and the Department of the Environment.

The Construction Press Ltd,
Lancaster, England.

A subsidiary company of Longman Group Ltd,
London. Associated companies, branches and
representatives throughout the world.

Published in the United States of America by
Longman Inc New York.

First published 1978

ISBN 0 904406 80 6

The opinions expressed in this book are the authors' and not necessarily those of the editor, publisher or the Construction Industry Conference Centre Ltd.

Printed in Great Britain at The Pitman Press, Bath

Contents

CONTRIBUTORS

Professor B. J. Brinkworth, MSc(Eng), PhD, MIMechE, MRAcS, MInstP, Director, Solar Energy Unit, Department of Mechanical Engineering and Energy Studies, University College, Cardiff.

Professor Peter Burberry, MSc(Arch), DipArch, ARIBA, Department of Building, University of Manchester, Institute of Science and Technology.

David A. Bütler, FRICS, Superintending Quantity Surveyor, Department of Health and Social Security.

Philip Caton, MA, PhD, FRMS, Head of Building Climatology Section, Meteorological Office, Bracknell, Berks.

James Dick, MA, BSc, FInstP, FCIBS, FIOB, Director, Building Research Establishment, Garston.

James P. Fyfe, FRICS, President of the Quantity Surveyors' Division of the Royal Institution of Chartered Surveyors, Partner, John Dasken and Purdie, Glasgow.

Professor G. Grenfell-Baines, OBE, DistTP, FRIBA, DipTP, FRTPI, Principal, Design Teaching Practice, Sheffield.

H. P. Johnston, DFH(Hons), CEng, MIEE, FCIBS, Deputy Chief Executive, Department of the Environment – Property Services Agency.

Julian Keable, FRIBA, AA Dipl, Founding Partner, Triad Architects Planners, London.

A. Leslie Longworth, MSc(Tech), PhD, CEng, FIMechE, MICE, FCIBS, MInstF, President, Chartered Institution of Building Services, Senior Lecturer, University of Manchester, Institute of Science and Technology.

J. Cleland McVeigh, MA, MSc, PhD, CEng, FIMechE, MIProdE, MCIBS, Head of Department of Mechanical and Production Engineering, Brighton Polytechnic, Organising Committee Member representing the International Solar Energy Society.

Professor John K. Page, BA, FIES, Chairman of the United Kingdom Section of the International Solar Energy Society, Professor of Building, University of Sheffield.

Geoffrey W. W. Pontin, DFH, CEng, MIEE, Director, Wind Energy Supply Company Ltd.

J. E. Randell, BSc, PhD, MCIBS, Senior Lecturer, University of Salford, Chairman, Organising Committee.

Kay J. Seymour-Walker, MA, CEng, MIMechE, RIBA, Head of Heat Pumps and Solar Energy Section, Building Research Station.

A. F. C. Sherratt, BSc, PhD, CEng, FIMechE, FCIBS, MInstR, Assistant Director and Dean of the Faculty of Architecture and Surveying, Thames Polytechnic, London, Organising Committee Member representing the Chartered Institution of Building Services.

C. V. Smith, MA, BSc, Head of Radiation Section, Meteorological Office, Bracknell, Berks.

R. Spicer, FRICS, Senior Partner, Northcroft, Neighbour and Nicholson, Organising Committee Member representing the Royal Institution of Chartered Surveyors.

Robert W. Todd, BSc(Eng), PhD, Technical Director, National Centre for Alternative Technology, Machynlleth.

Brian Trueman, BA(Manchester), Producer/Presenter, late of Granada Television Limited.

P. C. Venning, FRICS, Partner, Davis, Belfield and Everest, Organising Committee Member representing the Royal Institution of Chartered Surveyors.

D. F. Warne, BSc(Eng), ACGI, Head of Energy and Power Systems, Electrical Research Association Limited, Leatherhead, Surrey.

P. R. Warren, BA, PhD, Head of Heating and Ventilation Section, Building Research Establishment, Organising Committee Member representing the Department of the Environment and the Chartered Institution of Building Services.

Donald R. Wilson, BArch, RIBA, Architect and Lecturer, School of Architecture, University of Manchester.

Preface

Space and water heating in buildings account for a substantial proportion of primary energy consumption and pressure to conserve fossil fuels has caused the need to consume such large quantities of energy for heating to be questioned. Measures to reduce the heat requirements of buildings have in consequence become matters of keen interest to all concerned with the design and performance of buildings or with energy supplies.

There are a number of distinct if complementary, ways of reducing the fuel requirement of a building. In the first place, the building may be planned and constructed to make full use of the energy received by its envelope. Secondly, loss of heat through the fabric of the building can be reduced by the use of insulating materials. The heat requirement might be reduced further by the recovery of heat from air exhausted from the building. Ambient energy sources, which will provide a continuing supply of energy, can be used in place of fossil fuel to meet the heat requirement at least in part. In many of the more attractive proposals use is made of more than one source of energy. Combined use may be made, for example, of wind power and solar energy. With a heat pump solar energy can be used as a heat source. Ways of reducing the demand for fuel appear almost limitless in number and the range of alternatives available, including combinations, is bewildering.

The aim in this book is to review the principal possibilities regarded at the present time as practical. The practicability of an ambient energy scheme depends upon the availability of sufficient energy and the practicability upon the cost of the energy supplied. Proposals are considered in this light and specific experimental and design studies are reported. No review of the subject can be exhaustive but it is hoped that the information given and arguments presented will enable the reader to form a balanced view of the possibilities for energy savings in space and water heating in buildings.

J E Randell

ACKNOWLEDGEMENTS

The editor and technical organising committee wish to thank all the people who have participated in the arrangement and operation of the conference and in the production of this book.

A particular word of thanks is given to Mrs. Diana Bell and her colleagues at The Construction Industry Conference Centre for their efficiency in the organisation and administration.

1. Ambient Energy in the Context of Buildings

J. Dick

INTRODUCTION

The possible role of ambient energy in buildings has to be seen against the overall pattern of current usage of energy in buildings and against the ways in which this may change in the future. It is the purpose of this chapter to outline some of the relevant background information and generally set in perspective possible contributions involving more use of ambient energy in one form or another.

The importance of the building sector in the national scene can be demonstrated from available statistics on energy usage which show that some 40-50% of the total usage of primary energy can be attributed to building services. In setting the context, it is therefore useful to start by considering the overall energy consumption and then to develop in rather more detail the breakdown within the building sector as such, to cover the various types of building and their services and the energy demands involved.

But as well as considering current demands and how ambient energy might be developed to increase its contribution to meeting these demands, one also has to bear in mind that the demands themselves could well be modified by energy conservation measures: one can reduce the demand on traditional energy supplies by conservation measures as such or by increasing the use of ambient energy. Clearly in assessing the benefits of these two approaches, the same basic criteria for assessing cost-effectiveness should be used.

There are many possibilities for reducing the energy consumption of building services. A recent review (1) looked at a wide range of energy conserving measures and assessed both their maximum conservation potential and their cost-effectiveness. These range from the use of waste heat from thermal power stations in district heating to the use of the ambient energy from the sun and the wind. The review concluded that by undertaking the technically feasible options it should be possible to achieve an ultimate saving of over 15 per cent of annual consumption of primary energy by measures in building services which would not impair environmental standards. An important consideration in the quoted review was the extent to which the theoretical savings by, for example, improved appliance efficiency or increased insulation would be realised in practice, and this aspect is equally relevant in considering the possible greater use of ambient energy.

Thus one has to assess the demand for particular energy consuming services not only in terms of the quantities involved but also in terms of the distribution of demand in time; one has to consider not only the potential matching of energy supply and the apparent demand, but also the response of the user of the energy to changes in cost. It may be that the full potential benefit, whether from conservation measures or from increased use of ambient energy, is not realised in practice in that an individual occupant of a building may opt to take, or be forced to take, some of the benefit in improved standards. There are a number of cases where this has to be recognised, and an allowance made before assessing the economy likely to be achieved in practice.

Another important aspect of the current energy picture is the degree of uncertainty which surrounds any longer term prediction of the types of energy supply, their availability and of course their costs and subsequent pricing. The approach adopted by the Advisory Council for Research and Development on Fuel and Power in their discussion document 'Energy R & D in the United Kingdom' (2) was to examine the technology needs in a range of possible futures, in effect a series of scenarios based on assumptions such as low/high economic growth, national self-efficiency, high energy prices in the near future, a sudden price rise at some stage in the future. The underlying uncertainties about the future pattern have important implications for those examining energy in buildings and of course in particular those concerned with possible applications of ambient energy.

First of all, buildings have a long life compared with the rapidity of possible changes in the energy field and it is important to try to keep options open for a range of actions. Secondly, although the cost-effectiveness of certain measures may not justify action at the present costs and prices, changes in the future might substantially change the picture: it is therefore important that possible contributions are not discarded simply on current assessment but that the scope for technological development of the most promising items should continue to be explored.

ENERGY CONSUMPTION

In the UK buildings use more energy than any other single category of users such as "industry" or "transport". This fact, together with the long life of buildings, means that action on energy conservation and the planning of future energy supply are dominated by the crucial role of buildings.

When discussing the energy consumption it is important to distinguish between the consumption at or by the building itself (the net energy) and the consumption of the original raw material or primary fuel (the gross energy). The difference between the two can be great. For example, in the UK in 1972 only 27 per cent of the calorific value of the fuels entering the power stations arrived as electricity at the point of use. The "energy overheads" of the fuel industries will be discussed below.

The primary energy consumption of the UK in 1972 was 8.9×10^9 GJ* and energy delivered to final users was 6.1×10^9 GJ (3), the difference representing losses in conver-

Figure 1 **Gross energy input and net energy output of UK economy.**

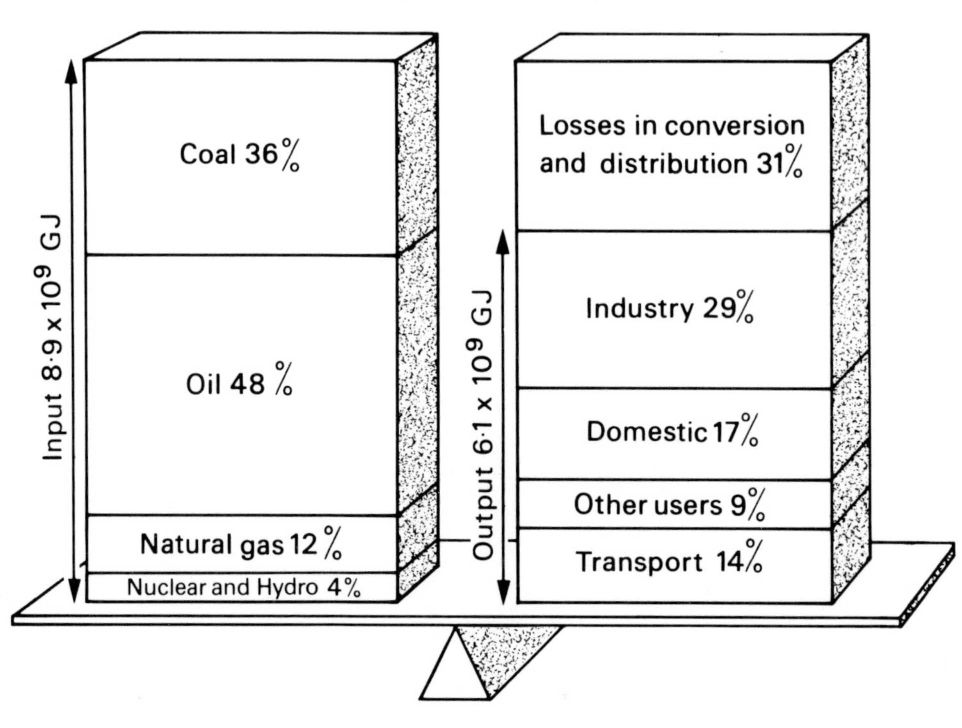

sion and distribution by the fuel producers (Fig 1). The relationship between the gross energy used and the net energy delivered for UK fuel industries is given in Table 1.

Table 1

Ratio of Gross Energy Input to Net Energy Delivered to Users: UK Fuel Industries, 1972

Coal	1.03
Oil	1.09
Natural Gas	1.07
Town Gas	1.42
Electricity	3.82
Other manufactured fuels	1.38

Fig 2 shows the percentage of the gross energy input attributable to different sectors, and is derived from Fig 1 by distributing the losses in conversion and distribution amongst the final users in proportion to their use of the various fuels; 29 per cent of the primary energy was consumed in housing, this being nearly twice as much as that consumed by the transport sector. Due to the lack of detail in the available statistics there is some uncertainty about the exact proportion of primary energy that is consumed in buildings. Certainly it will include all that consumed in the domestic sector, a large proportion of that in the "other users" sector and some from both the industry and transport sectors. The proportion is at least 40 per cent, and probably as much as half of the total primary energy consumed in the UK. The bulk of the building energy is accounted for by services rather than materials and construction, these energy uses in any case being included in the industry category.

Energy in Buildings

There were about 19 million domestic households in 1972, so the average net energy consumption in the home was about 81 GJ. This provided space heating, water heating, cooking, lighting and other sundry usages. Estimates based on field studies give the approximate usages shown in Table 2.

Table 2

Average Annual Domestic Energy Consumptions (GJ)

	Net	*Gross*	*Useful (approx)*
Space heating	52 (64%)	74	30
Water heating	17 (21%)	30	9
Cooking	7 (9%)	16	5
Lighting, TV, etc	5 (6%)	18	5
	81 (100%)	138	49

An important feature of the UK domestic sector is the relatively large usage of electricity. It is about twice the per capita usage of the original six EEC member countries. Belgium and Holland provide particularly good comparisons as they share with the UK a similar maritime climate; the per capita consumption of electricity in the domestic sector in the UK is about 2½ times that of Belgium and twice that of the Netherlands.

Clearly the largest consumption is for space heating, as would be expected. However, the consumption for domestic hot water of 21 per cent is not inconsiderable. The miscellaneous consumption for TV, lighting, etc is taken as 6 per cent of the total. There will of course be large variations from these mean estimates for particular dwellings, but the averages are helpful in providing the scale and pattern of usage.

There are certain difficulties in using existing statistics to assess energy use in non-domestic buildings but approximate estimates can be made. Fig 3 shows the estimated

* In this chapter the unit of energy employed is the joule J and its various multiples MJ = 10^6J, GJ = 10^9J. 1 GJ is approximatley equal to 278 kWh or 9.5 therms.

Figure 2 Gross energy used by final users.

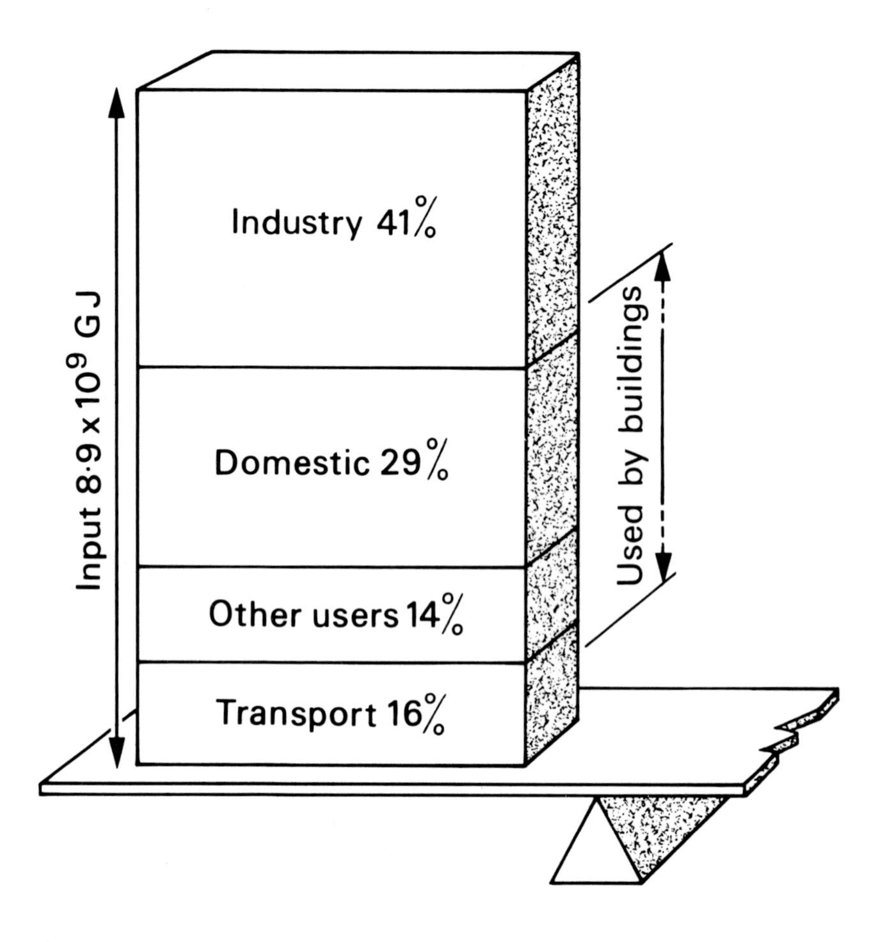

Figure 3 Estimated gross energy consumption in the categories public service and miscellaneous (derived from various tables in UK Energy Statistics 1973).

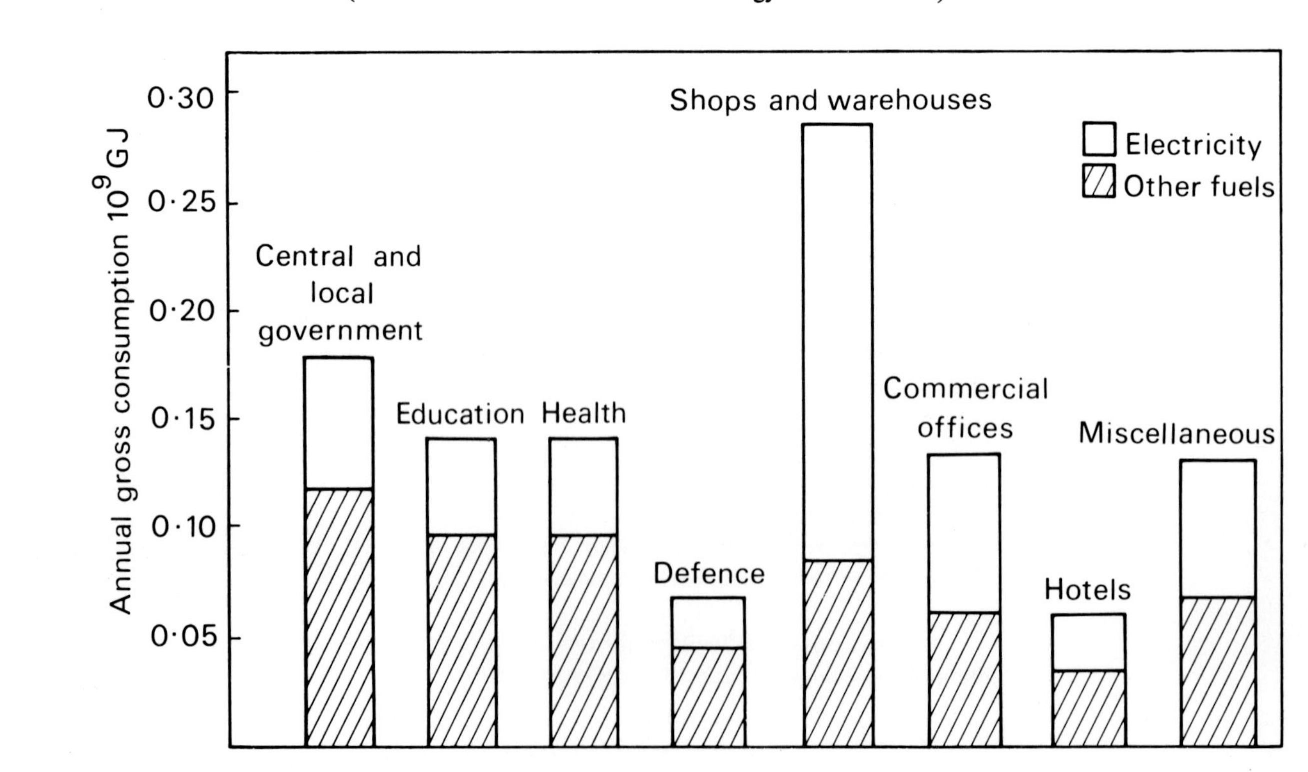

division of the gross energy consumed by the 'Other users' sector against eight categories. The pattern of energy use varies widely between different building types, and from the pattern of use in the domestic sector. Lighting plays a much more important role in the non-domestic sector. The lighting of offices and schools alone accounts for about one per cent of the national primary energy consumption. Lighting levels can also represent a dominant heat source and so incur a cooling load. Fig 4 illustrates the consequences for energy consumption, where the gross energy consumption of a shallow, naturally-lit, naturally-ventilated office is compared with that of a deep-plan air-conditioned office lit by traditional fluorescent light fittings to a level of 750 lux. The naturally-lit, naturally-ventilated office which maximises the use of ambient energy has a primary energy consumption almost one-third less than the air-conditioned, artificially-lit office.

The Energy User

The discussion so far has been in relation to the national energy consumption of particular building types but within these national consumptions there is a wide range of individual requirements which depend upon such factors as patterns of use and individual preferences. For example, a recent survey (4) of energy use in modern local authority three bedroom households in Scotland showed a mean net energy consumption of 42 GJ with the highest 5 per cent using about 5 times the consumption of the lowest 5 per cent.

This indicates that the savings from an energy conservation measure and hence its cost-effectiveness will vary greatly between apparently identical dwellings depending on the energy requirements of the occupants of the dwelling. Some dwellings maintain high temperature standards for long periods whereas others which are intermittently occupied are only heated for short periods each day. In addition to analysing energy conserving possibilities and carrying out controlled laboratory experiments there is clearly a need for field studies which span the wide range of individual variability. One example of such

Figure 4

Primary energy consumptions for three built-forms of tall offices
A naturally-ventilated, naturally-lit shallow plan
B air-conditioned, naturally-lit, shallow plan
C air-conditioned, artificially-lit at 750 lux, deep plan

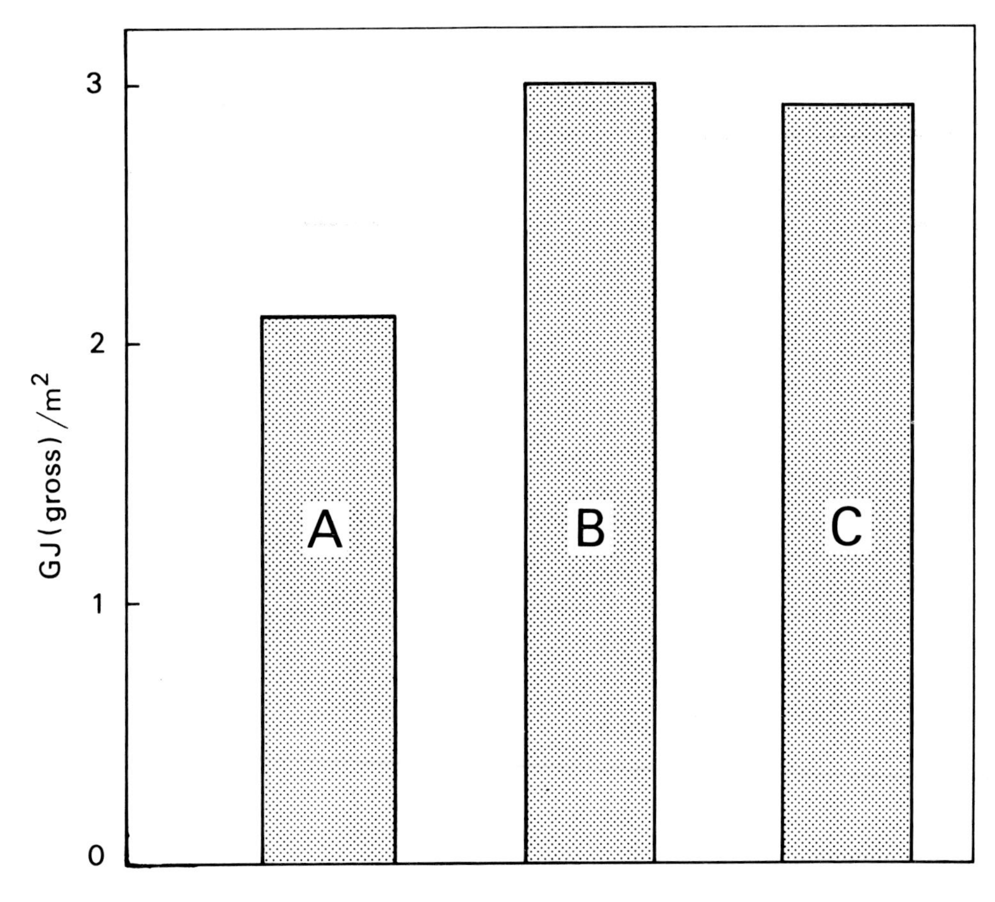

studies is shown in Fig 5. This shows the annual net energy consumption of 1000 local authority dwellings in Scotland as a function of the fabric transmittance (the thermal insulation) for electric and gas space heating. The net energy consumption of the electrically heated dwellings was about 80% of the gas heated dwellings. However, when the overheads shown in Table 1 are taken into account large savings in primary energy are shown to result from using gas. The performance in practice of energy conserving measures taking account of the behaviour of the user is discussed further later, particularly in relation to upgrading the thermal insulation of existing buildings.

Figure 5

BRE (Scottish Laboratory) results. The total annual net energy consumption per dwelling for gas and electric space heating against fabric transmittance.

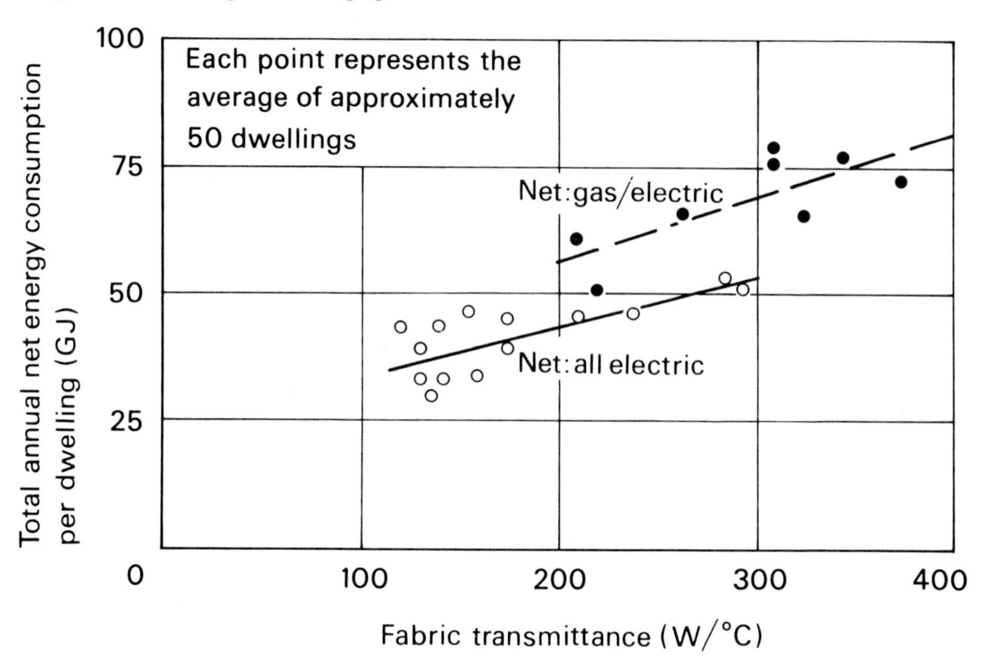

ENERGY CONSERVATION

The BRE study of energy conservation undertaken in 1975 and already quoted was essentially concerned with reviewing existing knowledge and possible implications for energy conservation measures, and also with defining areas where future research might be most profitable. Gaps were found in the information available but nevertheless an overall assessment was made which it was felt set the various problems in perspective. It was recognised however that the picture obtained might be modified to some extent by more detailed information on various aspects. The present conference provides the opportunity to bring together the experience of those involved in the ambient energy field and to fill in the picture on this aspect. It is probably therefore helpful to sketch out some of the main conclusions of the BRE review with a particular emphasis on the assessment made of ambient energy from the sun and wind. In this summary the emphasis will be on energy use in the most important category of buildings, namely housing.

The general conclusion from the review was that the ultimate realistic potential for gross energy savings in buildings is estimated to be about 15 per cent of the national consumption, which is about as much as is presently used by the transport sector. This total could only be achieved over a long period of time since it depends on the use of technologies which are more fitted to new constructions designed for the purpose. The estimated potential of possible measures applied in existing housing is about 6 per cent of total national consumption.

Combined Generation of Heat and Power

District heating schemes seldom offer substantial savings over individual fossil-fuelled appliances. They do, however, offer the opportunity for a combined generation of heat

and electrical power. It was estimated that about 10 per cent of the national primary energy consumption could be saved if all buildings could be heated by waste heat from power generation. There are however substantial problems in matching the load factors of the heat demand and the electricity demand which can adversely affect both the energy efficiency and economics. The feasibility and costs of combined heat and power for the UK is a complex issue not yet fully resolved.

Heat Pumps

Heat pumps when used for heating purposes extract heat (from the sun) via either an air or ground source and use mechanical energy to transfer the heat into the building. It seems technically feasible to design heat pumps with a heat energy output three times as much as the mechanical energy consumed. An electrical heat pump thus has a primary energy efficiency better than that of the best modern domestic boilers. It was estimated that the use of heat pumps to supply heat to buildings would save about 7 per cent of the national primary energy consumption. Heat pumps at present available for domestic use are not economic when compared with natural gas central heating systems, but there is scope for cost reductions when the designs are optimised for UK conditions. Their economics would be most favourable in applications where electricity was the only energy source available.

Better Primary Energy Efficiency

Electricity is a high-grade energy source in that it can provide motive power very efficiently even after allowing for power station losses. However, the demand for heating buildings is generally for low-grade heat and this can be provided with a better primary energy efficiency and usually more economically by the direct combustion of fossil fuels. In primary energy terms, gas is about twice as efficient as electricity for space heating.

Further savings can be made in fossil-fuelled systems by improvement of control and husbandry and in avoiding oversizing at the design stage. Savings of the order of 4 per cent in the national consumption of primary energy can probably be made by improvement in plant efficiencies. The cost-effectiveness of different measures varies.

Thermal Insulation

Improved thermal insulation saves energy by reducing the fuel consumption required for space heating and also enables smaller and cheaper heating plant to be used. The Building Regulations in force from January 1975 (5) constitute a significant advance in the mandatory requirements applicable to new housing. However, some improvement of thermal insulation of most of the existing housing is technically feasible. Not all of the theoretical savings may be realised in practice because the internal temperature is frequently found to rise after an improvement in thermal insulation. This can arise from a whole series of factors including poor system control, limitations imposed by the response of the building to intermittent heating, and to the occupant wishing to take some of the savings as improved thermal comfort. The estimated potential from up-grading the thermal insulation of existing housing is 3-4 per cent of the national energy consumption. Loft insulation and cavity fill are cost effective taken across the board which means that these measures not only pay but also improve the internal environment.

Solar Energy

The value of the direct solar gain through windows in winter is not always recognised as it might be, but of course it must be off-set by the heat losses due to the higher thermal transmittance involved. Designers intent on maximising the utilisation of solar gain in winter also have to bear in mind the possibility of overheating in summer. The optimisation of solar gains through windows and heat losses is now helped by the availability of simplified design aids (6).

Solar collectors can be used to provide space and water heating. On a theoretical basis, about 6 m^2 of collector is required to provide about half of the domestic hot water needs of an average dwelling; such an application is not cost-effective at present but future prospects are better. If universally applied where technically feasible 2% of the UK's primary energy consumption would be saved.

It is technically possible to include enough solar collectors and storage into a well insulated house to meet most of the demands for space and water heating. Such an experi-

Figure 6 Wind energy in the UK at an effective height of 10 m over an open site.

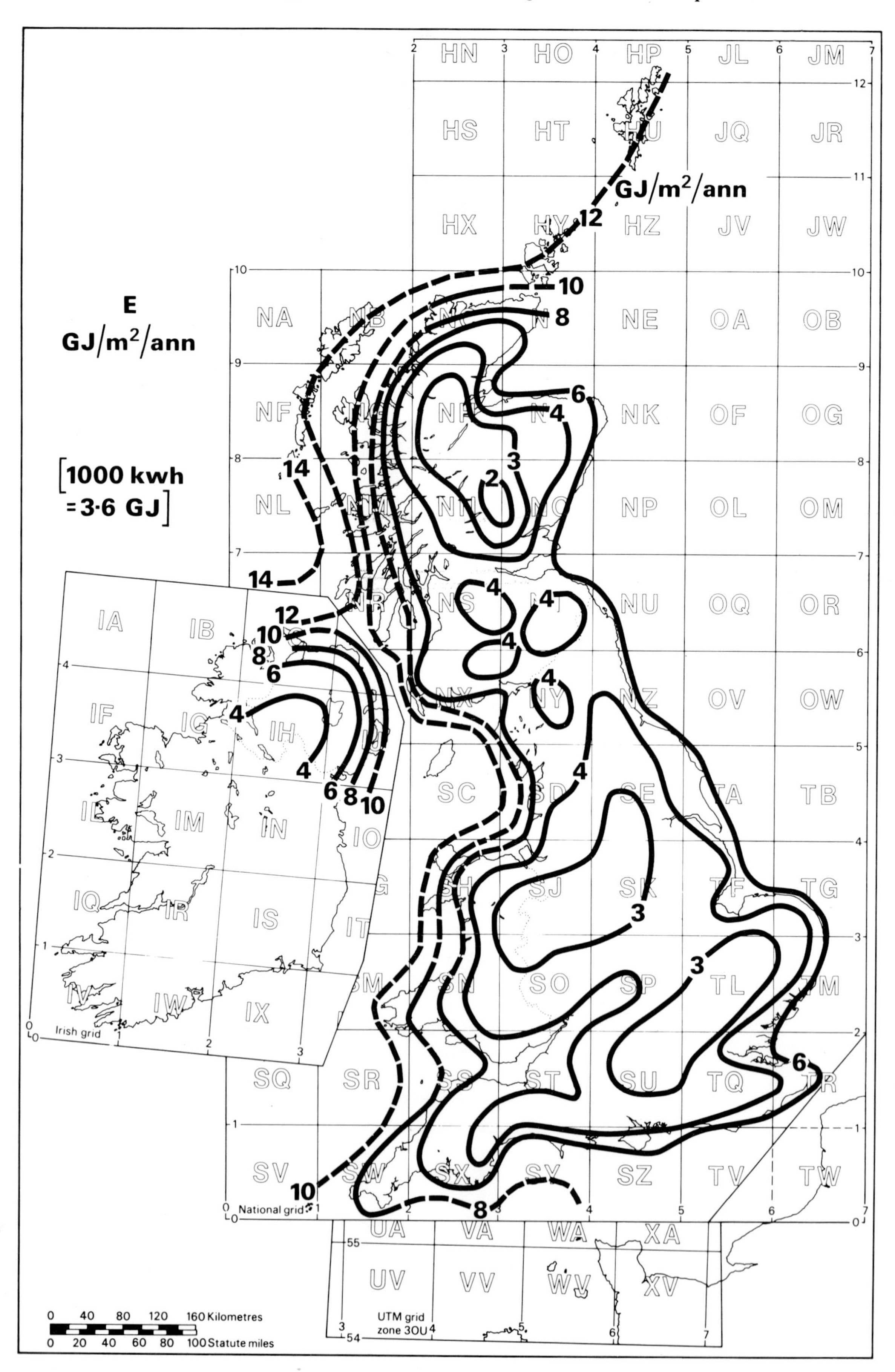

mental house is being constructed at BRE and will be described later. It is far from being cost-effective at present costs. However, in the long term solar energy for space heating could lead to substantial savings.

In order to put solar energy in perspective it is useful to calculate the areas of solar collectors involved. The annual average insolation per sq metre of horizontal surface of the UK is about 3.0 GJ. If we assume that the UK's annual requirements for low grade heat are about 5×10^9 GJ then solar collectors operating at an efficiency of 30% and covering an area of 6000 km^2 would be required – about the area of Norfolk. The associated heat storage would also be enormous although here economies of scale would apply.

Wind Energy

The UK climate makes the generation of wind energy favourable and wind energy is often available during the heating season. However, like solar energy associated storage or stand-by facilities are needed. Fig 6 shows the annual average wind energy at an effective height of 10 m over an open site. Since an aerogenerator can extract 33% of the wind's energy, in the least windy but open sites, a 10 m diameter rotor could in theory meet all the space and water heating requirements of the average dwelling.

On a per household basis this is not yet cost-effective. Also, in an urban situation only about half the open site's wind energy is available and it is clearly impractical to erect rotors with diameters exceeding 10 m on the roofs of houses. However, an analysis for a medium sized aerogenerator serving several houses in a windy district, shows that energy could be produced reasonably cheaply. Applications will also be very dependent on planning and aesthetic considerations and it is difficult to make realistic estimates of savings.

THE FUTURE

Although it has been seen that there are a considerable number of technical possibilities which can reduce the energy consumption of buildings without reducing standards, their relative importance in the future is still uncertain. As the decline in coal over the last 20 years has demonstrated, the predominant energy supply can alter in a time-span fairly short compared with the life-time of a building. The future is not likely to be any more certain. This emphasises the need for the designer to avoid decisions which severely limit the building users' future options for energy use. Keeping options open may not necessarily incur great expense. For example, space can be earmarked for possible storage of fossil fuels even if natural gas is intended to be used initially in a boiler installation.

For similar reasons, special attention should be paid at the time of construction to the thermal insulation of those building elements, for example walls, which would later prove difficult to up-grade. Good levels of thermal insulation in themselves open up a large number of options by lowering the overall heat demand to a level which can be met by one of the newer technical approaches such as solar collectors or heat pumps. It is important to consider not only the current implications for building design but also any provisions which should be made to accommodate future developments and changes and thus keep certain options open.

REFERENCES

1 *Energy Conservation: A study of consumption in buildings and possible means of saving energy in housing.* A Building Research Establishment Working Party Report CP 56/75, June 1975.

2 Advisory Council on Research and Development for Fuel and Power. *Energy R & D in the United Kingdom. A Discussion Document.* HMSO: 1976.

3 *United Kingdom Energy Statistics, 1973,* London HMSO.

4 Cornish, J.P. *The effect of thermal insulation on energy consumption in houses.* Energy Conservation in the Built Environment, 1976. Lancaster: Construction Press, 1976.

5 Statutory Instrument 1974, No 1944, *The Building (Second Amendment) Regulations 1974,* Order.

6 Milbank, N.O. *A new approach to predicting the thermal environment in buildings at the early design stage.* Building Research Establishment CP 2/74.

DISCUSSION

J. G. E. Williams (Eastern Electricity Board)

Mr Dick says gas is about twice as efficient as electricity for space heating and implies that using fossil fuels instead of electricity for home heating will help to conserve precious reserves of fossil fuels.

I question this view for two reasons. Firstly, Mr Dick considered the ratio of gross energy input to net energy delivered to users (Table 1) and this does not take account of the efficiency of utilisation at the point of use. It has been estimated that nearly a half of the fuel purchased is lost through plant inefficiency at the point of use. (Advisory Council on Research and Development for Fuel and Power. Energy R & D in the United Kingdom. A Discussion Document. HMSO: 1976). It is generally accepted that the efficiency of the conversion of electricity at its point of use is 100%, but for fossil fuel systems there is a great deal of evidence that the conversion efficiency over the heating season could be as low as 40%, much less, incidentally, than the figures suggested in other Chapters in this book (Keable, Chapter 6 and Seymour-Walker, Chapter 11). The second reason is that reserves of natural gas are in very short supply compared with coal. There may be over 400 years proven reserves of coal and, to quote an eminent authority, 'electricity is coal by wire'. My conclusion is that to use electricity for space heating is a way of using coal, much of it unsuitable for a domestic boiler, instead of gas, reserves of which are comparatively very small.

We need to keep fuel options open and all-electric buildings are one way of doing this. The most economical sources of energy can be accommodated by choice of fuel used at the power stations, avoiding installation of new heating plant in buildings. Furthermore, ambient energy supplies likely to be available in the future will most effectively enter the economy through the medium of electricity. The inference is that designing all-electric buildings is a way of conserving the most precious fuel reserves, and a practical way of keeping the future fuel options open for the nation.

J. B. Dick

In my paper I gave a comparison of consumption in some all-electric houses and comparable houses with gas space heating. The figures speak for themselves in terms of the rather lower net energy with the all-electric systems, but translated into gross terms, there is certainly a higher consumption with all-electric heating and also a higher cost. The market to a large extent will sort this out if realistic figures are fed back. I agree that gas from the North Sea should not be used without care – it is a scarce resource. I will express again caution about simply burning up all our fossil fuels which might well have a role as chemical feedstocks. Combined heat and power is a good example, related to electricity, of keeping options open. One might adopt rather different approaches in the higher density areas than in the peripheral ones. In the central areas, all the buildings might well be heated by district heating from a combined heat and power station; in the peripheral areas this would be less economic and a possible option would be to use heat pumps powered by electricity.

Professor B. J. Brinkworth (University College, Cardiff)

Mr Dick says that to attempt to meet the UK requirement for low temperature heating using solar energy at an efficiency of 30% would require a collection area of about 6000 km^2, roughly equivalent in area to Norfolk. As a warning against uncritical optimism about the use of solar energy I welcome it. An alternative view might be to calculate the area of buildings in the United Kingdom presented to solar radiation. This figure would probably be of the same order – certainly thousands of square kilometres.

Dr R. W. Todd (National Centre for Alternative Technology)

A direct boiler producing heat from fossil fuels has a theoretical maximum efficiency which approaches 100%, whereas the process of producing electricity is theoretically limited and efficiencies greater than 30 or 35% are unlikely. Even if heat pumps driven by electricity are used, the overall primary fuel efficiency is only equal to or may even be worse than that of a well designed direct boiler. Central heating boilers are only low in efficiency because of design.

Professor G. Grenfell-Baines (The Design Teaching Practice)

Overall efficiency of use of fossil fuels is what should be taken into account. Electricity may be 100% efficient in the home but it is less than 30% efficient overall. I would ask those with vested interests to have open minds when considering small local total energy schemes.

I agree completely with keeping options open. If the trend is towards capitalising ambient energy by means of electrical transformation, should we be designing with electrical energy in mind, even if at the moment we take advantage of high efficiency plant, some of which has a life of only 15 years anyway.

J. P. Fyfe (John Dansken & Purdie)

I would subscribe, wholly, to the view that in the discussion of a subject like ambient energy, vested interests must be ignored, else the discussion is pointless. People must be prepared to discuss openly and dissipate any vested interests.

Dr. G. Saluja (Robert Gordon's Institute of Technology Aberdeen)

Two different figures for the energy consumption of an average household are quoted. 81 GJ per annum for Great Britain as a whole and 42 GJ per annum for a particular group of households in Scotland. Can the difference be attributed to higher insulation standards in the Scottish households or are there other reasons to explain the difference?

J. B. Dick

The average worked out in the first part of the paper is for many millions of dwellings, which are not exactly comparable in size to the dwellings in Scotland. Neither are they necessarily comparable in climate, most of the dwellings being in England. I would not expect the figures to be very close.

2. Wind and Solar Radiation — Availability in the UK

P. Caton and C. V. Smith

INTRODUCTION

Wind and solar radiation are important sources of energy and may be used either as supplements to conventional supplies or as independent systems incorporating storage facilities. This chapter describes briefly the nature of the data concerning wind and solar radiation which are held by the Meteorological Office and, with the aid of a few simple examples, illustrates some of the ways in which these data may be analysed to provide information relevant to the design of practical systems.

SOLAR RADIATION

The amount of energy reaching the earth's surface depends both on the path length of the radiation through the atmosphere and on the amount of attenuating material. The path length of the direct solar beam varies with time of day, time of year and with latitude, and for a given place and time may be determined by simple geometry. Attenuating material is in the form either of clouds or of naturally occurring and man-made particulates, both of which forms fluctuate in time and in space in an inherently random manner.

Typical fluctuations in energy receipt at a UK station caused by clouds and aerosols are illustrated in Fig 1; the fluctuations are superimposed on changes caused by the diurnal variation in the path length. The scale of the fluctuations emphasises the need for measurements of radiation over a number of years and at a network of stations. They further imply that statements of radiation quantities for design and engineering purposes must necessarily be expressed in probabilistic terms and be based on the assumption that future years will fall within the sample represented by the recent past.

During its passage through the atmosphere the solar beam is subjected to scattering and absorption by material particles – atoms, molecules and their aggregates. These particles commonly redistribute the radiant energy taken up, sometimes in preferred directions, sometimes in all directions. The result is that, in addition to the direct solar beam, indirect diffuse radiation reaches the ground from the whole sky dome. This diffuse radiation is not isotropic, but has a maximum from the direction of the sun itself. The usual measurements are:–

a. – *the solar irradiance**, ie the flux of solar radiation on unit area of a horizontal surface from the complete sky dome. This is often called the global solar radiation and includes both direct and diffuse components.

b. – *the diffuse solar irradiance** on a horizontal surface; this is achieved by shading the receiving surface from the direct solar beam.

* According to the strict definitions, the 'irradiance' and its time integral the 'irradiation' refer respectively to the power and energy incident on a horizontal surface from the complete sky dome, whereas the terms 'radiance' and 'radiation' refer to the power and energy incident per unit solid angle from a particular direction. In this paper the word 'radiation' being in more general use, will be employed when speaking of the energy receipt from the whole sky dome.

Fig 1 **Typical variation of Global Solar Radiation on a partially cloudy day in summer.**

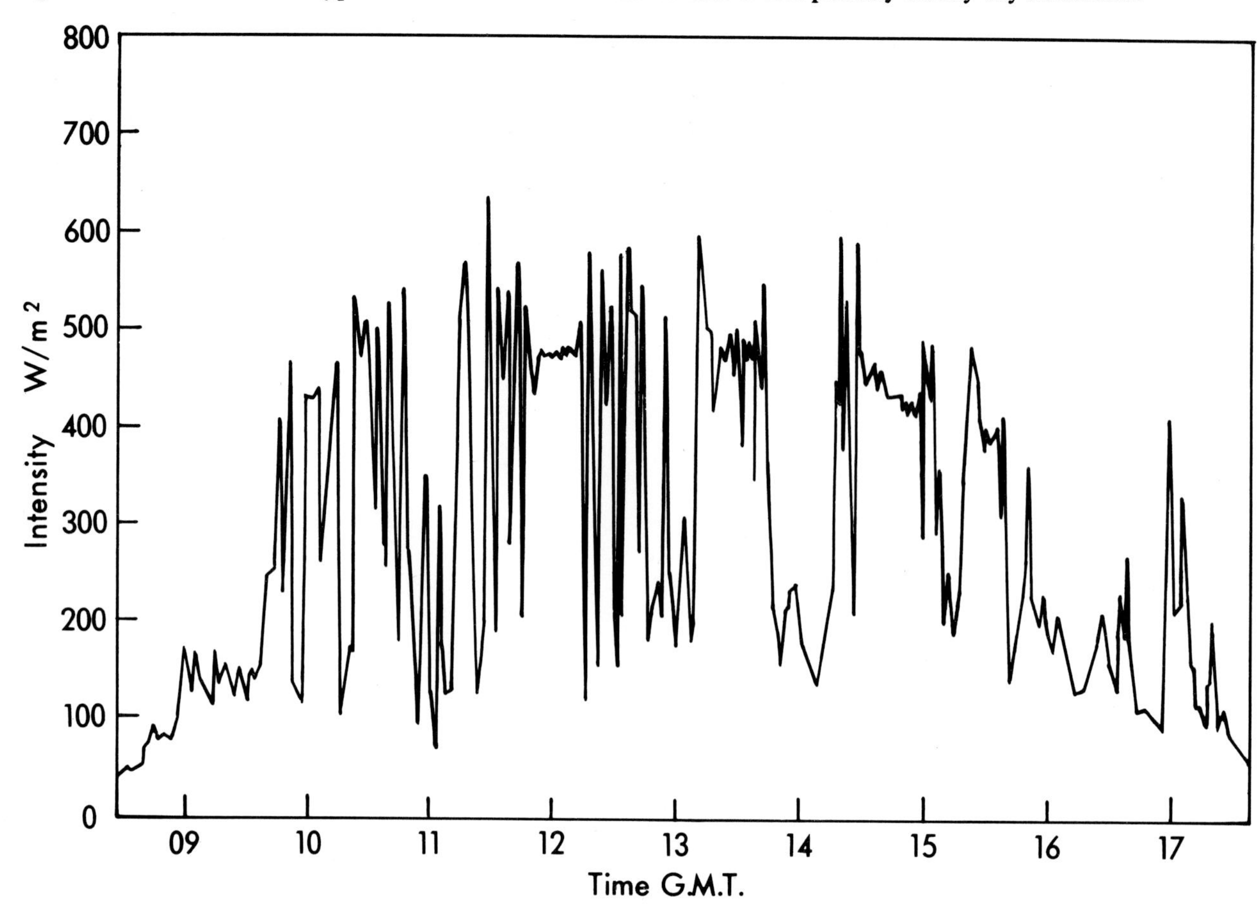

The basic field instrument makes use of the thermo-electric effect and consists of a collection of differential thermo-couples, aligned in series. The technique involves a sensitive electrical measurement and, as well as careful instrument design, careful maintenance and calibration procedures are essential to achieve worthwhile results. Data accuracy is estimated at around ± 5 per cent for individual hourly totals.

The SI unit of irradiance is the 'watt per square metre', but a convenient working unit, ten times larger, is the milliwatt per square centimetre. The SI unit of irradiation is the 'joule per square metre' (1 joule = 1 watt for 1 second). The period over which the radiant energy has been collected must be specified. The shortest period in general use is an hour, so for practical purposes the SI unit of irradiation is far too small; convenient working units are the milliwatt hour per square centimetre, the kilowatt hour per square metre and the megajoule per square metre (100 milliwatt hour/cm^2 = 1 kilowatt hour/m^2 = 3.6 MJ/m^2).

Hourly totals of global and diffuse radiation are available for the official climatological stations in Table 1. Preparations are in hand to double this network within a year from now. At these stations, data are logged each minute on paper tape and the archived data may be manipulated into any desired format; the basic data are available, at a nominal cost, either as computer printout or on magnetic tape, but the customer must pay for software preparation if this is needed to produce a special format. Logging on magnetic tape is to be introduced during 1977.

The Meteorological Office holds the National Reference Instrument, against which all radiation instruments used in the field should be calibrated. By offering such a service gratis, in exchange for the data collected, the Meteorological Office is able to acquire for the national data bank information from about 25 collaborating radiation stations maintained by Water Authorities, Experimental Farms, Universities, etc. The basic data pro-

Table 1 **Radiation stations providing hourly totals of global and diffuse radiation; also daily sunshine totals.**

Station	*Lat.*	*Long.*	*Elevation (m) above sea level*	*Date of commencement*	
Lerwick (W)	60 08N	01 11W	82	January	1952
Eskdalemuir (W)	55 19N	03 12W	242	January	1956
Aldergrove (W)	54 39N	06 13W	68	November	1968
Cambridge	52 13N	00 06E	23	January	1957*
Cardington (W)	52 06N	00 25W	29	January	1972
Aberporth (W)	52 08N	04 34W	133	July	1957
London W.C. (W)	51 31N	00 07W	77	January	1958
Kew (W)	51 28N	00 19W	5	January	1950
Bracknell (†)	51 23N	00 47W	73	February	1965
Jersey (gW)	49 13N	02 12W	83	January	1968
Aberdeen (gW)	57 10N	02 05W	35	June	1967
Dunstaffnage (gW)	56 28N	05 26W	3	April	1970
Dundee (gW)	56 27N	03 04W	30	July	1973
Hurley (g)	51 32N	00 49W	43	January	1970

W Mean hourly wind speeds available for all or part of period – see later sections.

* Ceased after December 1971.

† Measurements also of global solar radiation on vertical surfaces facing N, S, E and W, commencing January 1967.

g Measurements of global solar radiation only; no sunshine data.

vided by these collaborating stations are the daily totals of radiation on a horizontal surface, though some give additional information.

It should also be mentioned that almost 400 stations in the UK measure the duration of bright sunshine. By means of simple regression relationships, these observations may be used to interpolate radiation values between points in the solar radiation network.

The chief difficulty when attempting to use radiation data to meet engineering needs arises from the fact that the majority of basic measurements have been made on a horizontal surface. For averages and totals over periods longer than about 10 days, approximate relationships can be established between radiation measurements on a horizontal surface and estimates of the radiation falling on surfaces of other inclination and aspect. However, because the proportions of the direct and diffuse radiation components vary with the water vapour and aerosol content of the atmosphere, the transformation from horizontal measurements to vertical surface estimates for hourly totals of solar energy must be considered suspect unless account is taken of water vapour and turbidity at the specific place and time; this information is not normally available in sufficient detail.

Figs 2 and 3 show the average daily totals of global solar radiation over the UK for the months of June and December respectively. The maps are provisional and show only the broad-scale features of the distribution of radiation. They take account of the average values of sunshine duration as well as the radiation measurements. The principal factors which determine the pattern of isopleths are:–

(a) the general decrease in radiation levels with increasing latitude,
(b) the accentuation of radiation loss over a broad central area that covers the major conurbations, and
(c) the general decrease in average radiation totals in inland areas associated with the increase in cloud amount, particularly over high ground.

Table 2 shows for Kew, Aberporth, Eskdalemuir and Lerwick, the variation through the year of the average daily totais of global solar radiation on a horizontal surface. The table

Fig 2 **Global solar radiation – June.**

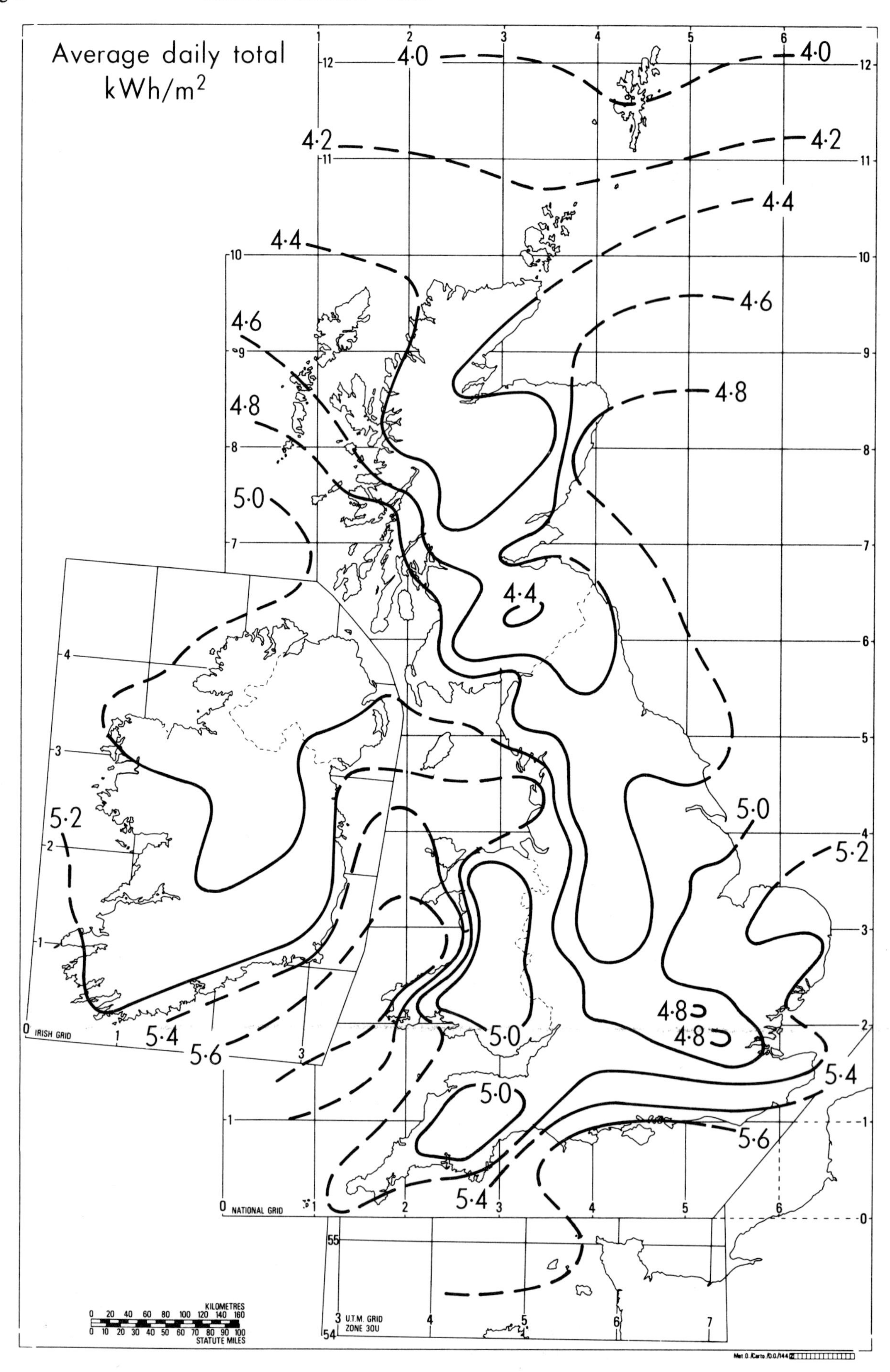

Fig 3 **Global solar radiation – December.**

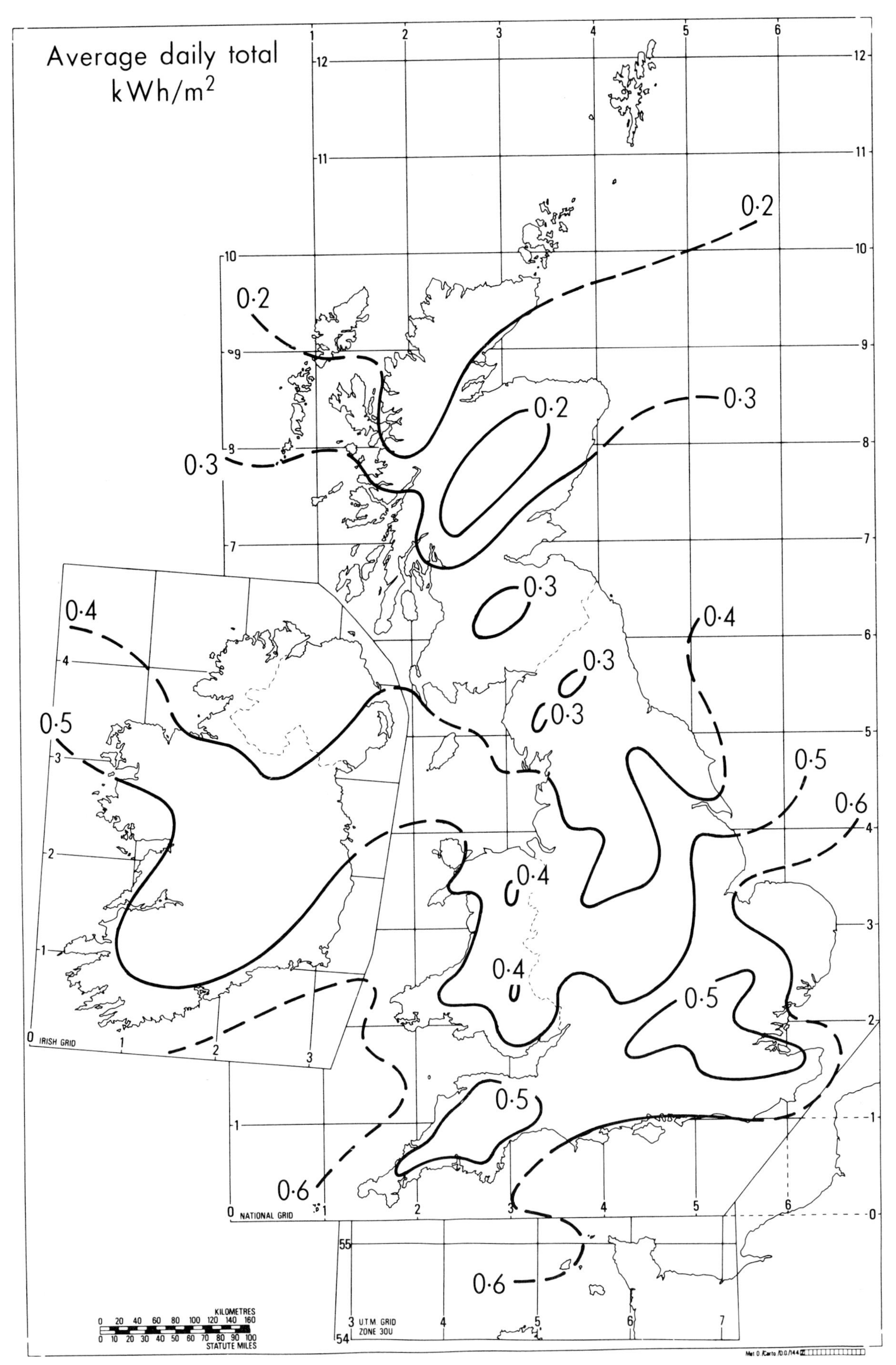

shows the change from winter to summer months to be by a factor of 8-10 in southern England increasing to about 20 in the north of Scotland. The mean daily solar energy estimated as available on a vertical south-facing wall is shown in Table 3 (1). In this case the change from winter to summer months is by a factor of 3½-4 in southern England increasing to 8-9 in the north of Scotland.

Table 2 **Mean daily totals of global solar radiation on a horizontal surface kWh/m² day, 1965-70.**

	Jan	*Feb*	*Mar*	*Apr*	*May*	*June*	*July*	*Aug*	*Sept*	*Oct*	*Nov*	*Dec*
Kew	0.59	1.15	2.24	3.23	4.32	5.02	4.45	3.69	2.70	1.61	0.83	0.48
Aberporth	0.66	1.40	2.64	3.96	4.69	5.58	5.04	4.19	2.94	1.71	0.83	0.54
Eskdalemuir	0.43	1.21	2.03	3.19	3.61	4.60	3.79	3.40	2.16	1.26	0.67	0.38
Lerwick	0.23	0.83	1.78	3.59	3.78	4.69	4.28	3.31	1.91	0.98	0.38	0.15

Table 3 **Mean daily solar energy estimated as available on a vertical south-facing plane kWh/m² day (Ref 1)**

	Jan	*Feb*	*Mar*	*Apr*	*May*	*June*	*July*	*Aug*	*Sept*	*Oct*	*Nov*	*Dec*
Kew	0.9	1.5	2.4	2.5	2.8	3.0	2.7	2.6	2.4	1.9	1.3	0.8
Lerwick	0.4	1.5	2.1	3.0	2.7	3.1	2.9	2.6	1.8	1.4	0.7	0.4

There is a wide variation in the total of global solar radiation on individual days. Table 4 gives, for Aberporth, an analysis of the percentage levels of the frequency distribution of

Fig 4 **Diurnal variation of Global Radiation on a horizontal surface. Each percentile line joins the appropriate values in the frequency distribution of hourly average values of radiation intensity at Kew, based on measurements in June 1964-73.**

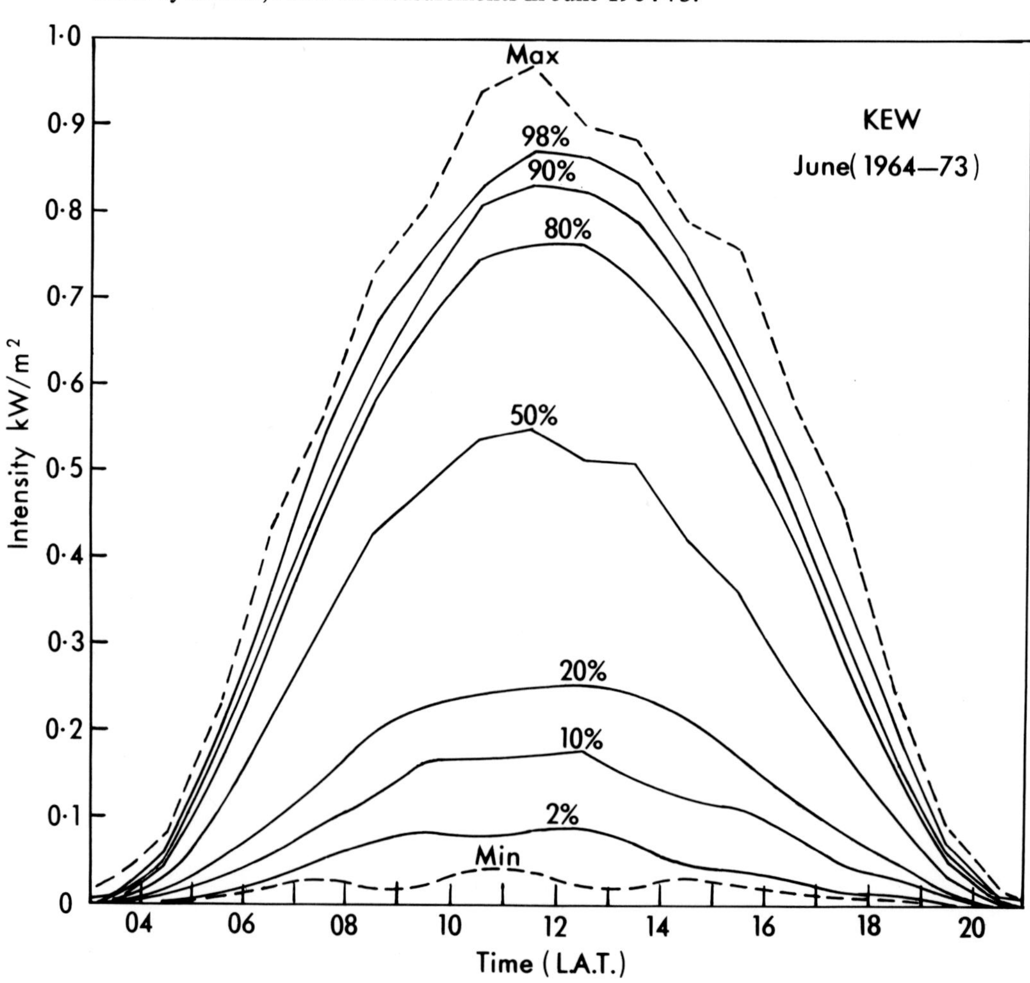

daily totals on a horizontal surface, based on measurements 1959-72. Based on a similar analysis of hourly totals. Fig 4 illustrates the diurnal variation of global radiation through days in June, based on measurements at Kew 1964-73. Of course, on any actual day, the radiation totals for individual hours will vary above and below the particular percentile level.

Table 4

Percentage levels of the frequency distribution of daily totals of global solar radiation on a horizontal surface at Aberporth 1959-72.
Unit: Kilo Watt-hours per square metre

Percent Level	*Jan*	*Feb*	*Mar*	*Apr*	*May*	*June*	*July*	*Aug*	*Sept*	*Oct*	*Nov*	*Dec*
Lowest	0.09	0.14	0.38	0.36	0.58	0.64	0.61	0.26	0.45	0.14	0.08	0.04
1	0.13	0.27	0.55	0.82	1.02	0.83	1.23	0.63	0.49	0.29	0.13	0.09
5	0.21	0.40	0.84	1.32	1.85	2.05	2.00	1.67	1.09	0.49	0.24	0.15
10	0.25	0.54	1.06	1.64	2.42	2.61	2.55	1.98	1.44	0.68	0.31	0.20
20	0.33	0.73	1.44	2.25	3.19	3.58	3.43	2.68	1.90	0.96	0.41	0.30
50	0.62	1.27	2.55	3.90	5.13	5.83	5.21	4.38	3.18	1.73	0.80	0.53
80	1.05	2.02	3.72	5.46	6.88	7.80	6.93	5.74	4.39	2.49	1.22	0.86
90	1.25	2.38	4.21	6.10	7.47	8.09	7.51	6.40	4.70	2.91	1.46	1.00
95	1.43	2.62	4.60	6.46	7.83	8.31	7.83	6.83	5.02	3.16	1.67	1.07
99	1.72	2.91	5.28	6.90	8.21	8.67	8.40	7.29	5.73	3.69	1.96	1.16
Highest	2.00	3.16	5.38	7.13	8.86	8.97	8.65	7.68	6.03	3.96	2.13	1.22
No of obs	432	396	430	413	425	420	430	431	419	433	420	433

Table 5 gives, for Aberporth, the number of days within a 10 year period, subdivided into seasons, for which the daily total of global radiation lay below the limits shown. The partition of these days into spells is also indicated. The spell count was strictly prescribed by the days at the beginning and end of each season, and there was no contribution from days outside even though spells may have continued into an adjacent season. The spells shown in any one row, that is below one limit for the daily total, are independent; spells below higher levels naturally include days that have been counted at lower thresholds. One obvious conclusion from the data is the probability of long spells in winter with little significant input of solar radiation. Even in May, June and July, days with only moderate solar energy input, eg those with daily totals of global energy less than 3 kWh/m^2 form 17 per cent of the population and may occur in spells of two, three or even four days.

WIND

Currently there are about 145 'official' anemograph stations in the United Kingdom. The network includes a substantial number of stations run by co-operating authorities, which provide data to the Meteorological Office in exchange for regular inspection and maintenance by technical staff. Since January 1973 all stations have tabulated for each clock hour the mean hourly speed and the maximum gust. In earlier years the mean hourly speed was tabulated for each hour but only the daily maximum gust was recorded. The data are quality controlled and those since January 1970 are available on magnetic tape either on an individual month all-stations basis or on an individual station all-years basis.

The ratio of the maximum gust speed to the mean speed for individual hours at an effective height of 10 metres is referred to as the gust ratio G. Mean values of G have been derived using the 21 864 hourly observations between January 1973 and June 1975, subdivided according to hourly mean direction and hourly mean speed; they provide a useful characteristic of the surface roughness surrounding each anemograph site and their use in producing maps of hourly mean wind speed corresponding to standard roughness conditions is described by Caton (2). He also explains the use of maps of wind speed corresponding to standard roughness when attempting to predict the wind climate of particular sites for which data are not available.

Table 5 **Radiation spells at Aberporth 1964-73**

Upper limit daily total of global radiation kWh/m²	*Total of days*	*Spell Duration (Days)* 1	2	3	4	5	6	7	8	9	10	*11 and over*	*Max. spell length days*
Winter Nov-Dec-Jan													
1	737	27	20	10	13	7	7	5	1	4	3	21	63
2	918	–	–	1	–	–	–	–	–	–	–	11	92
(Total No. of days in analysis 919)													
Spring Feb-Mar-Apr													
1	133	48	17	8	3	3	–	–	–	–	–	–	5
2	384	61	26	11	4	5	3	2	–	3	–	9	22
3	565	54	27	14	10	2	2	1	2	–	–	12	36
4	697	30	16	11	11	5	4	1	3	1	–	10	53
5	801	19	8	5	8	4	1	2	–	–	–	13	78
6	859	5	2	4	–	1	0	2	1	–	1	13	86
7	891	–	1	–	–	–	–	–	–	–	–	10	90
(Total No. of days in analysis 892)													
Summer May-Jun-Jul													
1	9	7	1	–	–	–	–	–	–	–	–	–	2
2	59	44	6	1	–	–	–	–	–	–	–	–	3
3	156	87	22	7	1	–	–	–	–	–	–	–	4
4	295	97	36	26	6	1	2	1	–	–	–	–	7
5	433	81	47	30	16	4	3	3	2	2	–	1	11
6	585	58	31	23	19	10	5	9	9	2	–	7	14
7	741	16	20	9	9	8	12	6	11	6	5	16	28
8	878	1	3	2	3	1	–	1	–	–	1	31	92
(Total No. of days in analysis 920)													
Autumn Aug-Sep-Oct													
1	81	55	11	–	1	–	–	–	–	–	–	–	4
2	283	84	30	9	9	–	3	1	2	–	1	2	14
3	515	75	31	8	3	2	4	3	1	1	–	11	38
4	671	43	29	12	7	3	1	2	1	–	2	12	47
5	813	18	12	8	3	5	4	3	1	5	–	11	75
6	877	5	6	6	2	1	5	1	4	–	1	11	76
7	914	1	–	–	1	–	–	–	–	1	–	10	92
(Total No. of days in analysis 919)													

The power of a wind V acting on an area A is $\frac{1}{2}\rho AV^3$ where ρ is the air density. If the speed V is maintained for time T the total energy in the wind is $\frac{1}{2}\rho AV^3T$; if V is measured in knots (1 knot = 0.15 m/s), A in m², T in hours, the expression becomes 0.83×10^{-4} AV^3T kWh (or 3.00×10^{-4} AV^3T MJ) In practical terms, if V = 28 knots the energy E = 1.82 AT kWh; if V = 20 knots the energy E = 0.66 AT kWh. Currently the average efficiency of a windmill rotor and associated electrical generator is about 40 per cent (3); accordingly the maximum output energy at a speed of 20 knots is about 0.27 AT kWh. Since the speed V always varies with time, the product V^3T should properly be regarded as $\int V^3(t)dt$ where the interval dt is determined by the response time of the aerogenerator to fluctuating speed. The mean hourly value of the cube of a fluctuating speed always exceeds the cube of the mean hourly speed, but the excess is likely to be only a few per cent and will not be considered further in this chapter.

Fig 5

Typical wind speed – duration and wind power-duration curves for a site in the United Kingdom. The shaded area under the power-duration curve represents the 'available energy' using an aero-generator with rated speed V_R, cut-in speed V_C and furling speed V_F.

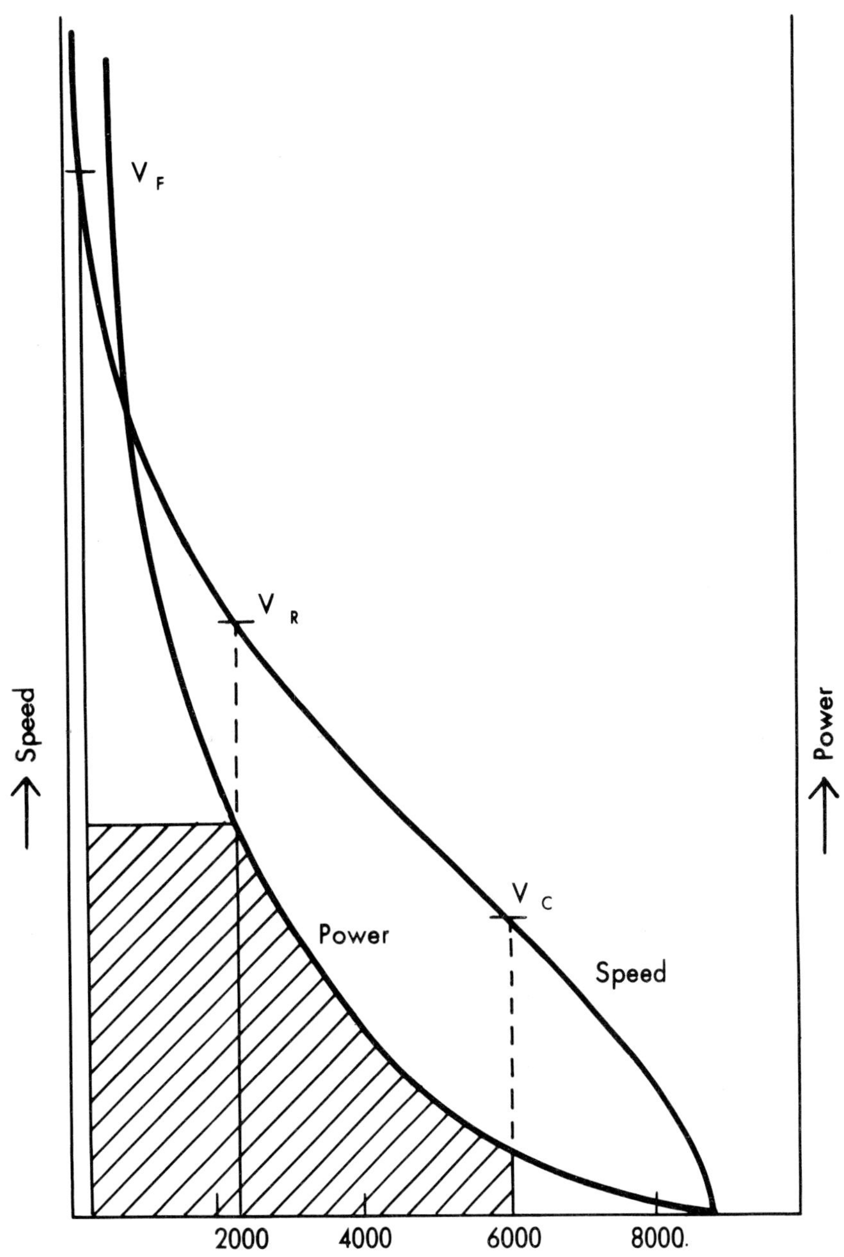

Fig 5 shows (a) a typical velocity-duration curve for a site in the UK, and (b) the derived power-duration curve. The shape of the curves varies relatively little between sites but the units of speed and power depend appreciably on height above ground, surface roughness, topography and location. Maps showing the hourly mean wind speed exceeded for various percentages of the time have also been published (2).

Aerogenerators have a rated wind speed V_R corresponding to maximum output, and a cut-in speed V_C which is normally about 0.5 V_R . They must be shut down at a furling speed V_F to prevent damage. The 'available energy' over a long period of time is thus represented by the shaded area under the power-duration curve in Fig 5. Hopkins (private communication) has evaluated the 'available energy' (in megawatt hours per square metre per year) at seven anemograph sites in the UK for various values of the rated windpseed V_R, (V_C was assumed equal to 0.5 V_R in each case). His results are shown in Fig 6.

If the objective is to extract maximum energy over the year as a whole, the graphs suggest that there exists an optimum rated speed for each site. This optimum rated speed V_R is plotted in Fig 7a against the average wind speed $\overline{V}$ for the site and it appears that a simple relationship exists $\hat{V}_R = 2.3\ \overline{V}$. The average wind speed for a site may be obtained using a simple cup-counter anemometer to measure the 'run' of wind over a period, or may be estimated from the maps and instructions in Climatological Memorandum No 79 (2). Golding (4) has given examples of hill-top sites having particularly high average wind speeds.

Fig 6 **The available wind energy at selected sites as a function of rated wind speed.**

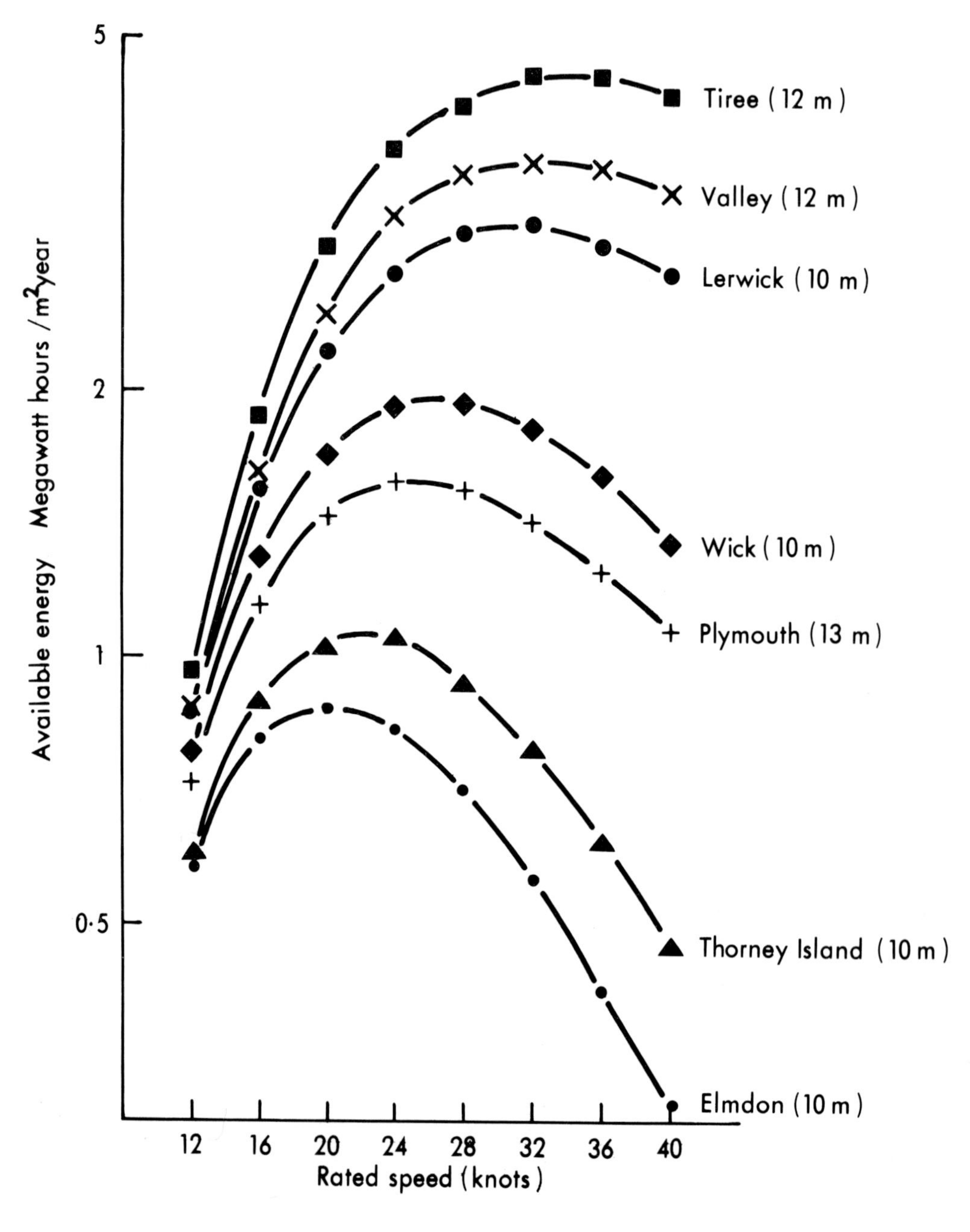

Fig 7b, which is derived from Figs 7 and 7a, is a plot of the available energy at optimum rated speed against average wind speed. This graph confirms the cube relationship in the form $\hat{E} = 1.3\ (\overline{V}kn)^3$ kilowatt hours per square metre per year; ($\hat{E} = 9.5\ (\overline{V})^3$ kWh/m^2 year when $\overline{V}$ is in m/s). The relationship is applied in Fig 8 to produce a map of available wind

Fig 7 (a) Optimum rated speed as a function of mean wind speed.
(b) Available energy at optimum rated speed as a function of mean wind speed.

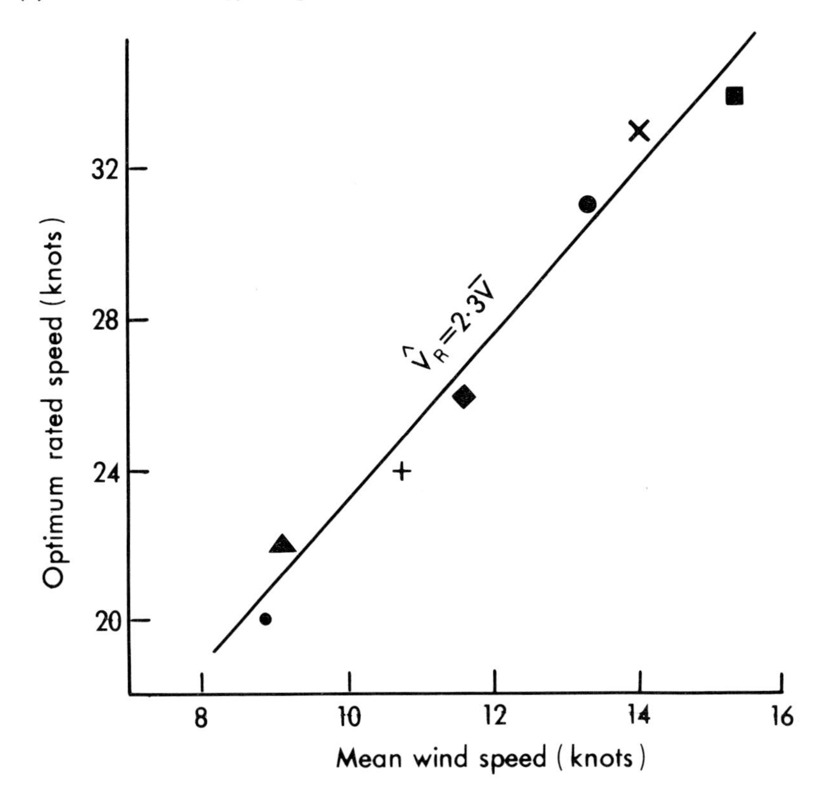

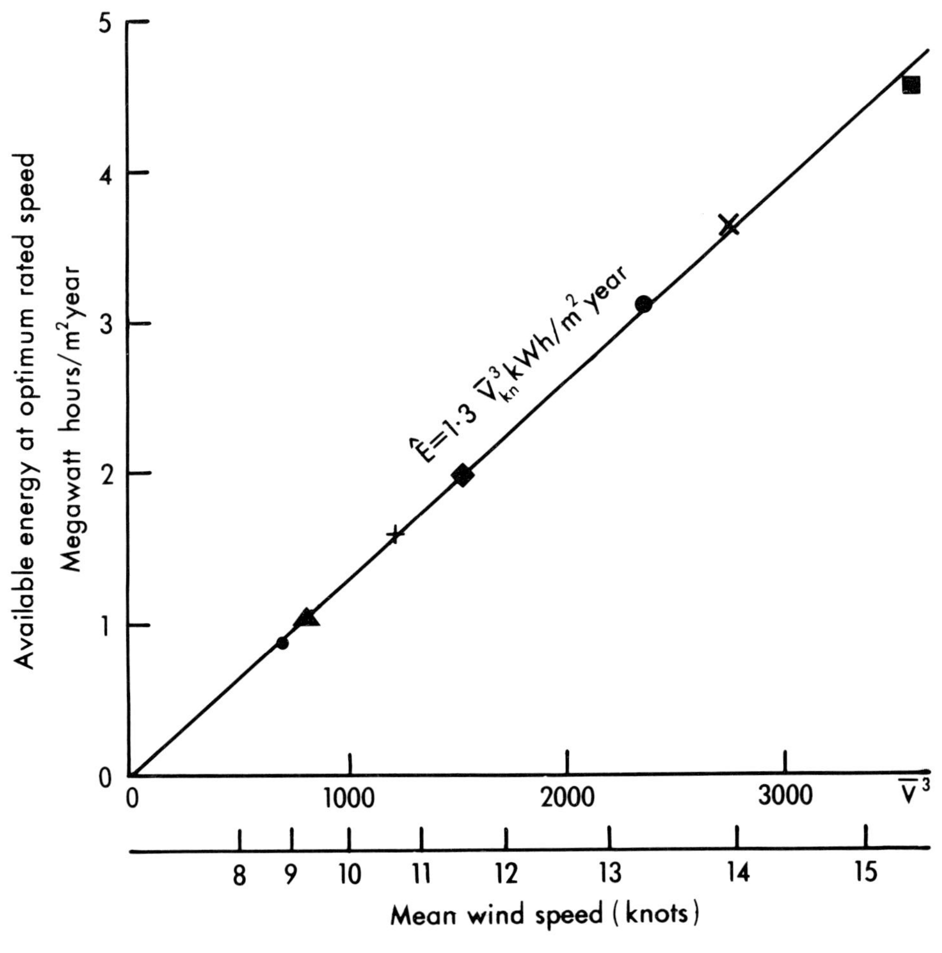

Fig 8 **Available wind energy at optimum rated speed.**

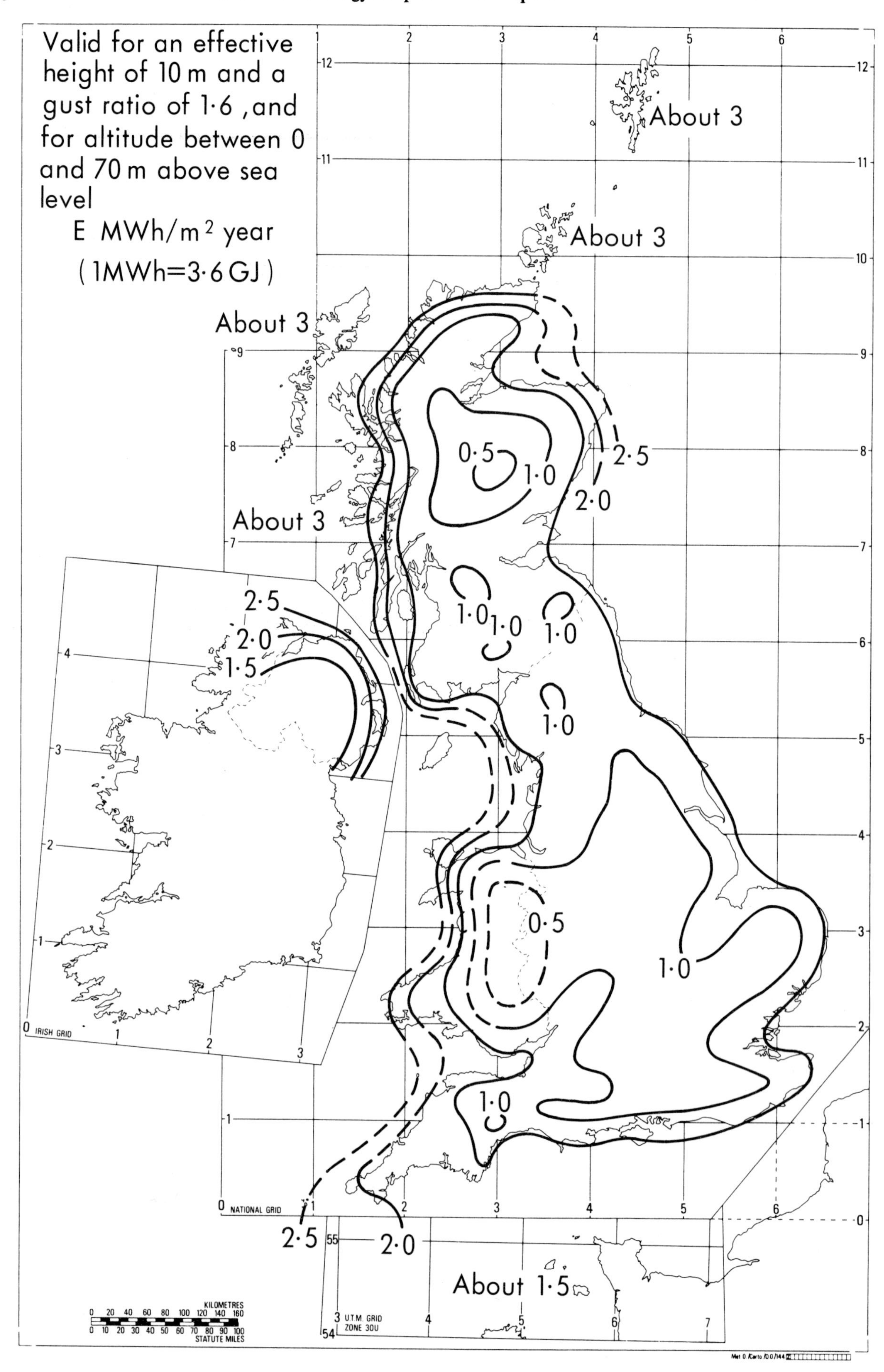

energy at optimum rated speed over the United Kingdom. The map is applicable to a height of 10 metres above open level terrain (gust ratio 1.60) and for altitudes between 0 and 70 metres above mean sea level.

The available energy will be greater than the map value at open coastal sites and over open high ground, in particular above smooth hill-tops. The available energy will be less than the map value over well-wooded parkland, and over towns and cities. Estimates of the adjustment factor to take account of surface roughness may be obtained by evaluating the cube of the appropriate values in Table 1 of Climatological Memorandum No 79 (2). These adjustment factors are very significant; at a coastal site the factor by which the map value should be multiplied falls within the range 1.2 to 1.4 (the higher figure applying when a very high proportion of winds blow from the sea) whereas in the centre of a large city the factor may be only about 0.33

The wind speed above the summit of a hill is very dependent on the height, shape and smoothness of the hill. Golding (4) has given examples of hills about 200 m high where

Fig 9

Spells with hourly mean wind speed below threshold values.

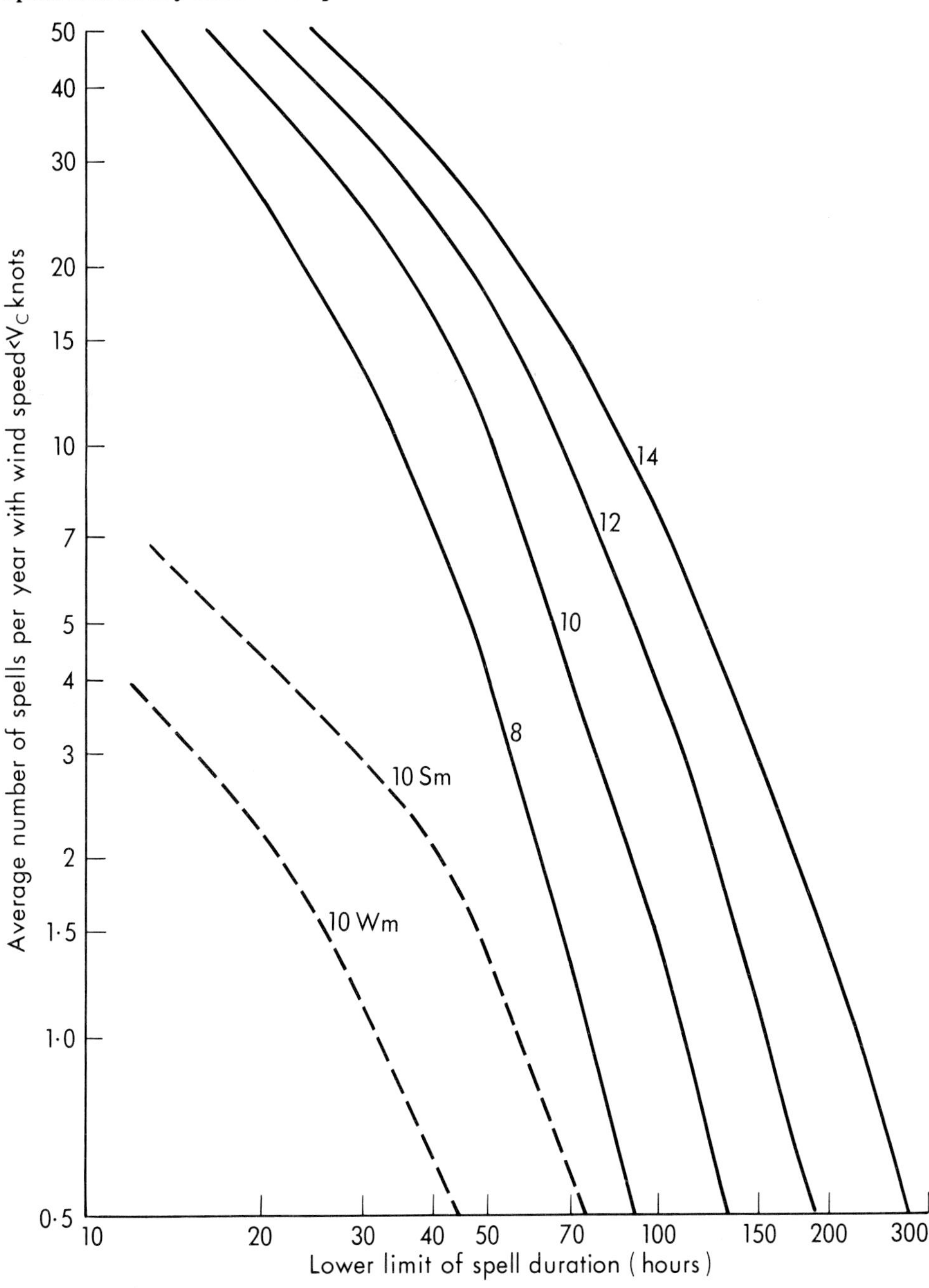

the average wind speed at the summit is about 60 per cent higher than at 10 m height above surrounding low ground. This corresponds to an increase in wind energy by a factor of 4. These special situations apart, it may often be possible, through careful attention to the siting of an aerogenerator, to achieve a 5 to 10 per cent increase in average wind speed compared with the standard meteorological site in flat open terrain; this modest increase in wind speed leads to an increase in wind energy by a factor between 1.16 and 1.33.

An important limitation to the use of wind energy is the intermittent nature of the energy source. The continuous curves in Fig 9 show, for Aberporth, the average number and duration of spells per year with mean hourly speed below threshold values of 14, 12, 10 and 8 knots. If an aerogenerator at 10 m height at Aberporth is designed with V_R = 28 knots and V_C = 14 knots to give maximum energy output over the year as a whole, the number of spells with zero output is quite substantial. However, if lower values of V_R and V_C can be accepted (at some sacrifice to total energy output over the year – see Fig 6), more continuous performance can be achieved.

Given that $V_C = 0.5\ V_R$ it appears that design compromises are called for depending on the major objective of the installation, ie maximum output or improved continuity of operation. However, if it were possible to design a generator for which V_C equals say $0.4\ V_R$, the system could be significantly more effective. For the same value of V_R, the total energy output would increase slightly but a more important fact is that there would be a marked reduction in the periods with no output; alternatively, for the same value of V_C, and the same periods with no output, there would be a substantial increase in the total energy output over the year.

The dashed curve labelled 10 Sm in Fig 9 shows the average number of spells per month with wind speed less than 10 knots for each of the summer months May to September, and the curve labelled 10 Wm shows the same information for the winter months November to March; April and October are, on average, transitional months and their curves fall between 10 Sm and 10 Wm. It will be seen that the majority of spells with mean hourly speed below 10 knots occur during the summer half-year.

Fig 10 **Average monthly totals of output power from solar collector and wind generator at Aberporth.**

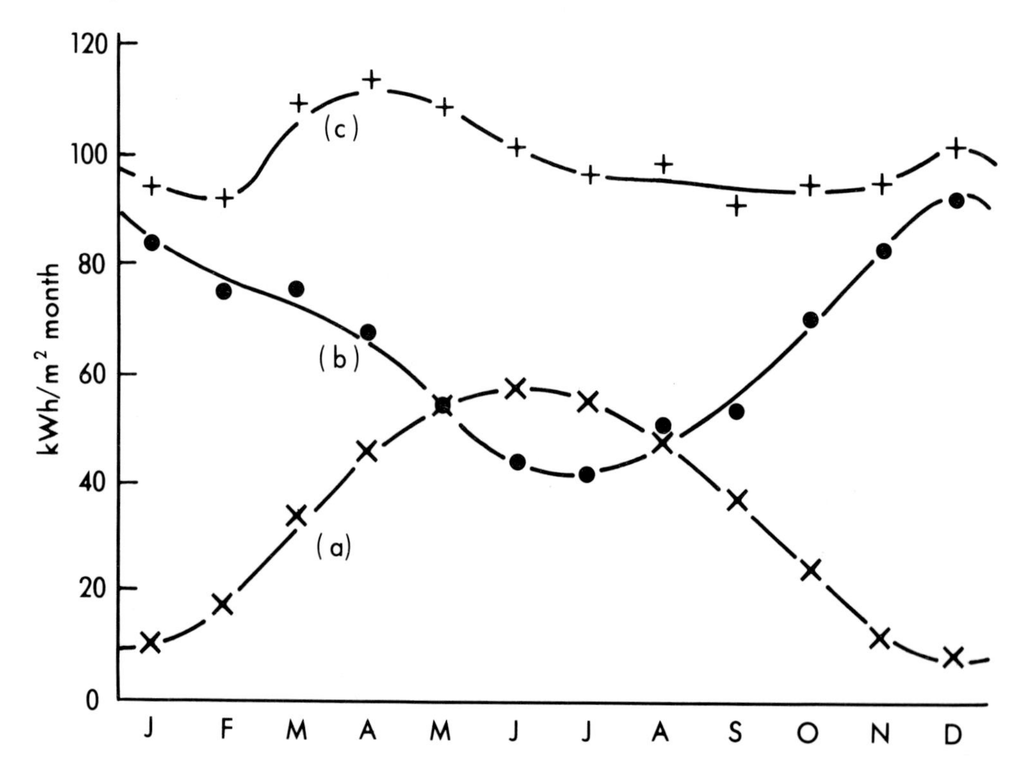

THE COMPLEMENTARY USE OF WIND AND SOLAR ENERGY

It will have been noticed that solar energy is both more plentiful and more continuously available during the summer months, whereas wind energy is greater in amount and less intermittent during the winter months. Fig 10 shows, for Aberporth the average monthly totals of output power expected from:

(a) a solar collector of area 1 m^2, on a south-facing roof at elevation 30 degrees, with efficiency 35% (5);

(b) an aerogenerator of area 1 m^2, at 10 m above ground, with V_R = 20 knots, V_C = 10 knots (V_F = 60 knots) and efficiency 40%;

(c) the sum of (a) and (b).

The monthly values of (c) are remarkably uniform throughout the year. Should the requirements include additional energy for heating in the winter months, the area of the aerogenerator may be increased relative to that of the solar collector, or the aerogenerator may be sited higher than 10 m above ground or in some other situation of enhanced average wind speed, or it may be designed with a higher rated wind speed.

With both radiation and wind data on magnetic tape, it is possible to carry out quite detailed analyses. For example, computer programmes could be written to investigate the simultaneous potential of solar and wind energy on a day to day or hour to hour basis. These programmes could be designed to take account of the performance characteristics of a proposed installation. Analysis of hourly values of wind and radiation is possible only for those stations marked with a W in Table 1, but conclusions drawn from these stations might possibly be adapted to other sites through the use of appropriate weighting factors depending on standardised wind speed, surface roughness and sunshine duration.

As an example of what can be done, a simple analysis has been made of days when only small amounts of energy would be available from (a) an aerogenerator alone, (b) a solar collector alone, and (c) an aerogenerator and a solar collector. This analysis is based on daily mean wind speed and daily sunshine duration and may be made for about 80 stations in the UK. Table 6 shows, for Aberporth, the percentage of days occurring in sequences of 1 day or more, 2 days or more, and 3 days or more with (a) daily mean wind speed less than 10 knots, (b) daily sunshine duration less than 1 hour in summer, 2 hours in spring and autumn, and 6 hours in winter, and (c) daily mean wind speed less than 10 knots and, simultaneously, daily sunshine duration less than the amounts above. The threshold of sunshine duration was varied with season in an attempt to simulate a threshold of approximately 3 kWh/m^2 of solar radiant energy incident on a south-facing roof at elevation 30

Table 6 **Spells with minimal energy from wind and/or solar radiation at Aberporth, Nov 1960 - Oct 1970**

Season		*Mean daily wind speed knots*	*Sunshine Duration hours*	*Per cent of days in Sequences of duration*			*Maximum Sequence Length days*
				≥ 1 day	*≥ 2 days*	*≥ 3 days*	
Winter	a	< 10	–	22.1	15.8	11.0	8
Nov-Dec-Jan	b	–	< 6	90.9	90.2	89.3	71
	c	< 10	< 6	17.5	10.9	5.2	5
Spring	a	< 10	–	28.6	21.6	18.7	12
Feb-Mar-Apr	b	–	< 2	39.9	29.4	17.8	8
	c	< 10	< 2	9.5	4.3	0.7	3
Summer	a	< 10	–	47.7	42.6	37.1	15
May-Jun-Jul	b	–	< 1	17.0	7.4	2.4	7
	c	< 10	< 1	6.9	2.5	0.8	4
Autumn	a	< 10	–	36.0	31.3	25.4	15
Aug-Sep-Oct	b	–	< 2	35.8	23.7	16.1	8
	c	< 10	< 2	10.3	5.0	2.8	4

degrees; however this threshold could not be achieved in winter, when even 6 hours of bright sunshine was estimated to correspond on average to only about 1.8 kWh/m^2 of energy on a south-facing roof at Aberporth.

The results show the percentages corresponding to circumstance (c) to be very much lower than those for circumstances (a) and (b), particularly for spells of 2 days or more. They demonstrate that with both systems available (even if used only as alternatives) the number of days with little or no output is substantially lower than with either system alone. This would reduce the storage requirement in any installation designed to be independent of conventional energy supplies. It should be noted that Aberporth is a favoured site for the collection of solar radiant energy and for the use of aerogenerators; particular sites must be considered on their merits.

ACKNOWLEDGEMENTS

We are indebted to our colleagues in the Meteorological Office for many helpful suggestions and in particular to J S Hopkins of the Climatological Services branch whose work has been reproduced in Figs 6 and 7a. The contents of this chapter are Crown Copyright.

REFERENCES

1. International Solar Energy Society, United Kingdom Section. *Solar Energy, its potential in the United Kingdom and the current need to expand research and development.* September 1974.
2. Caton, P G F. Climatological Memorandum No 79. *Maps of hourly mean wind speed over the United Kingdom.* Meteorological Office, 1976.
3. Rayment, R. *Wind Energy in the UK.* London, IHVE, Build Serv Eng, **44** 1976, pp 63-69. (Also available as Building Research Establishment, Current Paper No 59/76).
4. Golding, E W. *The generation of electricity by wind power.* London: E and F N Spon, Ltd, 1976.
5. Courtney, R G. *An appraisal of solar water heating in the UK.* Building Research Establishment, Current Paper No 7/76.

DISCUSSION

J. D. L. Harrison (Energy Technology Support Unit, Harwell)

Does Dr Caton think that the availability of solar energy in the major urban areas might be overestimated from present insolation distribution maps? Does he think that there is a need for the collection of more data from the urban areas, where the majority of people live, if reasonably accurate estimates of the energy available through passive and active solar collection systems are to be made?

Dr. P. Caton

The list of Meteorological Office radiation stations given in Table 1 includes Kew, in the South-West suburbs of London and the London Weather Centre. The average totals of global solar radiation at the London Weather Centre are about 5% below those at Kew. It is a matter for designers, engineers and others concerned with buildings to decide whether additional measurements in urban areas are required and, if so, for joint discussion on the ways and means of providing them.

Dr. R.W. Todd (National Centre for Alternative Technology)

Fig 9 shows spells with winds below certain threshold values. The problems of storage associated with wind powered systems are of particular interest. Is there a simple method of taking into consideration how close together calm spells occur? A spell of four days with wind speeds of less than say 9 knots may be followed by a windy day, but if another calm spell of similar duration is then experienced, this will have implications on storage requirements.

Dr. P. Caton

The problem is complicated and there is no simple method. It is not only a question of evaluating the spell duration and of providing sufficient storage capacity to cover the lean period but also of ensuring the energy is in the store at the start of the lean period. The latter depends on the wind speed over the preceding months, including lean periods. It is a problem suitable for computer analysis, which could be based on the data we have but for the work to be done there would have to be seen to be a requirement for the product.

J. P. Fyfe (John Dansken & Purdie)

Before funds are made available to carry out work, there has to be seen to be a requirement for it, an end product and a purpose. Is money available for work on ambient energy when the outcome is doubtful?

J. Keable (Triad Architects and Planners)

The question of storage and the prediction of storage requirements is one of the most important aspects of ambient energy. The storage requirement has a tremendous effect on the cost benefit of a particular scheme. No one individual in the field can afford to pay for the sort of programme that is required for predicting this. How do we pay a little each?

Dr. P. Caton

It is Government policy that we should charge for staff and computer time in running programmes, and also for the writing of special programmes. Where new investigations or programmes are necessary it may be advantageous if these are commissioned through a Professional Association or Government Department. An example of this approach is a service of 'degree day' information which is available to all who require it through the Energy Conservation Unit of the Department of Energy.

Professor G. Grenfell-Baines (The Design Teaching Practice)

The suggestion that the ideal site for collecting wind energy is a rounded hill top leaves out the problems which arise with wooded areas. I have heard it said that the best site for collecting solar energy is a treeless desert, but as a layman I wondered if I would like to live in a treeless desert to take advantage of solar energy.

J. Owen Lewis (University College, Dublin)

The 'Standard Year' based on the meteorological data seems an attractive idea. I understand the Danes have published a 'Reference Year' and there are others (including the Irish Meteorological Service) preparing similar typical sets of climatic data. What is Dr Caton's view on the use of such simulations?

Dr. P. Caton

What is the purpose of the 'Standard Year'? Our weather patterns are almost infinitely variable and it is not possible to select one year as a representative year for all purposes. The danger is that a representative year selected for one locality and for one particular purpose will be used for some other locality or some other purpose for which it is not appropriate. Energy consumption is likely to depend principally on temperature, wind and solar radiation. An average year for temperature is not necessarily an average year for wind, or for solar radiation. Further, a year with average temperature overall may not include an average number of cool and warm spells. The recommended procedure is to use continuous data for 15 or 20 years, which we can provide for a network of stations. There is a group in this country developing a representative year for heat energy calculations, but we have consistently stressed the difficulties.

P. Bond (Peter Bond Associates)

My contribution is to highlight the key role of conservation techniques applied to the fabric of the building. If we are considering whether or not sufficient solar or wind energy is available, in any particular area, to provide the energy needs of a normal dwelling, it may be found that there is inadequate energy available. But by the application of conservation techniques, it is possible to reduce the energy needed for the dwelling to the point where the ambient energy which is available is quite adequate.

By the application of conservation techniques, the first advantage is a reduction of energy need, but a further advantage is that capital expenditure involved in the equipment required to capture the ambient energy is scaled down correspondingly.

On a national level I have made studies which demonstrate the possible savings in energy consumption that can result from conservation techniques and the figures look good. If a costing factor on the savings made by scaling down the hardware needed to develop ambient energy is then added, the result represents a vast reduction in the national budget.

Considered thus, I believe conservation techniques are of paramount significance in any debate on the use of ambient energy.

3. Ambient Energy—Economic Appraisal

D. A. Butler

OBJECTIVES AND FACTORS INFLUENCING AN APPRAISAL

Background

During recent years the world has doubled its demand for energy every 10 years. Coal, gas and oil, the traditional fuels, have seen us through the energy needs of this period of industrial expansion. However these fuels are obtained from limited stores of energy which will some day become exhausted. Pressures are already developing for alternative resources to be found.

Sun, sea and wind are sources of energy offering continuous supplies for a limitless period of time. Although the crisis may not be imminent and thrifty use of the fossil fuels should ensure their availability for a few centuries yet, there is an inevitability surrounding the need to develop these alternative ambient energy resources. The only real questioning concerns the economics and timing.

In this chapter problems and cost factors are introduced in order to demonstrate techniques.

Research

As the cost of traditional sources of energy rises so alternative resources become increasingly attractive propositions. Many imaginative projects are being researched.

Sun In France there is the 50 metre diameter mirror concentrating the sun's rays to produce a temperature of over 2700°C. In Russia moving mirrors have been used to provide electricity.

Sea Another project envisages making use of the flow of water through the Straits of Gibraltar to produce electricity. Again in France tidal power, utilising the influence of the moon rather than the sun, has been used to operate a power station in Brittany. It has been estimated that if tidal energy could be harnessed this would produce more than the world's requirements for energy. However the anticipated cost is a major deterrent to development although recently the Parliamentary Under-Secretary, Alex Eadie, stated that "the Government is committed to the whole idea of wave energy" and that "money itself cannot accelerate development that can be made".

Research is a major drain on capital resources. The Department of Energy have so far expended £1 million on wave energy research. On completion of the current research programme more capital will be needed to carry the work through to a conclusion.

Wind Windmills have been used since early times as a source of energy but their use diminished as alternative and more reliable resources became available. Nonetheless it has been estimated that the energy of the wind could provide many times over the world's energy requirements. Unfortunately present windmill designs have severe limitations. It would take far too many of them to make an effective and significant contribution and the impact on the visual environment would be considerable.

In Britain it is estimated by the Central Electricity Generating Board (CEGB) that to replace one 2000 MW power station it would need 4800 windmills each 100 metres high.

The impact on the visual environment would be far greater than that of the power station they replaced and would probably outweigh any benefits gained by other aspects of the environment.

Pricing Policies

National and international politics have a considerable influence on the relative economics of energy supplies.

National Many economists argue that the price of energy should be fixed by the cost of production from the most expensive source if the most efficient use is to be made of those resources. This is not an argument that meets with ready acceptance and national policy is normally to allow each fuel to find its own price level and enter the market in competition with alternative fuels.

Such policies at present tend to act to the advantage of the consumer of natural or North Sea gas. However, the recent example of oil provides an example of the disadvantages of this policy. In the 1960's oil was a cheap form of energy against which other fuels could barely compete. Coal particularly suffered a period of depression and mines were run-down. Following the OPEC decision to increase prices coal suddenly became more competitive in price and the demand increased. The economic cost comparisons of alternative fuels were in many cases completely reversed.

Recent developments of North Sea oil and natural gas have made it certain that Britain will be self-sufficient in energy until the end of this century. Self-sufficiency can however conceal dangers not least that of complacency.

International. The 1967 White Paper on Energy Policy made the basic assumption that "regular supply of oil at competitive prices will continue to be available". This was totally upset when in 1973 the OPEC countries decided to limit supplies and increased the price several-fold.

No longer were the OPEC countries prepared to sell their oil at a price that reflected only the cost of production plus a small profit. They appreciated the strength of their position as the major suppliers of an internationally vital commodity. International demand and local supply combined to provide an opportunity to extract a much higher profit and almost overnight the price of oil increased by around 400 per cent. World energy demands are ever increasing and provide ample opportunities for the pronouncement of national and international decisions of considerable effect and consequence.

Sources

One problem met with when attempting to assess the cost of a fuel is that there is no single cost figure for production.

Coal. Coal from surface mines is a lot less costly to obtain than coal from deep mines.

Oil. Oil from land based wells is a lot cheaper to extract than that taken from wells located in the North Sea. One source quotes costs of well development ranging from Index 100 in Middle East to Index 3000 in North Sea with USA Indexed at 1500.

Hydro. Hydro-power may be economically produced in mountainous regions but prohibitively expensive elsewhere.

Gas. Natural gas is appreciably cheaper to obtain in Britain than other energy resources but this reflects the fact that the gas is being obtained from the shallow areas of the North Sea. As these areas are exhausted the gas will have to be obtained from more expensive wells in deeper seas and then the cost will escalate.

General. There are already strong indications that by the end of the century the more accessible supplies of fossil fuels will be seriously depleted. Moving to the less accessible sources will increase the cost of production and price to the consumer. The consequential effect on alternative fuel cost comparisons calculated on present day values is difficult to forecast.

Forecasting

Demand. Forecasting future energy requirements is a complex exercise plagued with uncertainty. Factors range from national economic performance to the degree of success achieved in the efforts to conserve and make better use of energy.

Current estimates vary between an increase of 50 per cent and 100 per cent of present demand by the year 2000. The 1967 White Paper on Energy Policy underestimated the 1970 primary energy demand by 6 per cent and overestimated the 1975 demand by 9 per cent.

Costs. It has been shown how income from the supply of energy to consumers can be affected by political decisions but future income is also dependent on the amount consumed. Development of production in excess of demand is expensive and a waste of capital resources.

Social Factors and Benefits

Social benefits offered by the various fuels are too often ignored or overlooked. Coal production is labour intensive and although the mechanisation programme of recent years has improved conditions the work remains arduous and potentially dangerous. Recent claims by miners for early retirement and a shortening of the number of years spent working undergound at the coal-face are a measure of the cost to the individual of working in such an environment. It is not difficult to appreciate the benefit to the individual if he can improve his working conditions and there may be a benefit to the community if the consequence is to reduce demands on the health services for example. Equally if these demands have not been considered when making a forecast the consequences could be serious for the decision-maker.

Environmental Factors and Benefits

Environmental benefits attaching to various fuels is another important factor. The visual effect of windmills has already been mentioned but atmospheric pollution is another aspect of effect on the environment.

Air pollution from the burning of coal is well-known and over 400 years ago the burning of sea-coal in London was forbidden to avoid the effects of the sulphurous fumes. Electricity, considered by many to be clean fuel, contributes to pollution as the generation of electricity usually involves the burning of fossil fuels.

Environmental pollution also occurs with hydro-electric schemes which often occupy large areas of land to allow the formation of artificial lakes. Atomic power creates the serious problem of radio-active waste disposal.

Solar energy and other ambient energy resources can mostly be developed with very limited impact on the environment.

Appraisal

An economic appraisal to be effective must consider all those factors that are in any way related to the subject of the appraisal. The factors referred to in the early part of this chapter can be collected into 3 main categories:–

1. Benefits – social and environmental.
2. Costs – production and consumption.
3. Forecasts – international and national policies and demand.

CONSIDERATION AND APPLICATION OF TECHNIQUES

There are many techniques available but on the principle that – simplicity is better understood than complexity – it is proposed to examine 3 widely applied and accepted techniques. Producing an appraisal is normally a 3 step operation with each step employing a different technique.

(i) Establishing costs and benefits
– Cost benefit analysis and traditional estimating.

(ii) Presentation in terms of present day costs
– Discounting.
(iii) Confidence limitations risk and uncertainty
– Probability analysis.

Techniques

Cost Benefit Analysis. Although this is an investment appraisal technique normally applied to public sector investments it is a technique that is being applied more often to a wide variety of issues in the private sector.

The main purpose is to introduce a discipline to the consideration of relevant facts by collecting all available data in quantative terms commensurable as possible to provide a comprehensive view of an issue.

In the private sector financial outlays and returns may provide sufficient criteria for commercial judgements. In the public sector however the strictly financial appraisal is inadequate. Judgements on energy projects with a public sector interest must surely take account of all relevant factors particularly where these concern the environment, the nation's energy resources and trading balances. Identified factors range from different fuels to the forecasts of future demand and encompass political decisions, capital investment, research costs and social benefits.

Some factors can be directly linked to expense or income in financial items and can be dealt with by normal estimating procedures. Other factors such as the positive or negative benefit to the environment can only be expressed in financial terms by the application of subjective judgements.

Factors naturally expressed in financial terms present few difficulties whereas those with no direct financial link can be a problem. How, for example, is the effect on the environment of building a power station to be valued in money terms, or the effect of constructing 4800 windmills each 100 metres high. It is certainly much easier to accept that such factors should be given some weight when proposals are being considered and if alternatives are to be properly compared.

Often it is necessary to exercise judgement to put a value to a factor making the best use of collective opinions. Such opinions may be gathered from experts in a particular field. In the above example possibly land valuation experts.

Discounting. Recent years have seen the technique of discounting money flows being adopted for a wide range of activities.

To achieve the objective of comparability and to provide an assessment of economic viability it is necessary for all inputs to an appraisal to be expressed to a common base. It is essential that benefits and costs are measured in money terms and that these measurements are presented in a given time stream. It is usual to express in terms of value at the present point in time, ie now.

Measured money values of benefits and income and costs and expenditure will naturally occur at various points in time and, as in the case of energy projects, often over a period of many years. It is convenient to divide time into equal discrete periods of a year and assume that all values that flow in a period can be treated as if they all occurred simultaneously at the end of that period. In this way cash flows associated with a project can be established in terms of a dated set of values, as is illustrated in the example given at the top of the next page.

Few would accept that the difference between expenditure and income and benefits of £83 million represented the true profit or value of the total project. Nor in the private sector would the difference between expenditure and income of £61 million be considered as the measure of the economic viability in commercial terms.

Years	Research	Capital	Operating	Income	Benefits
	COSTS/EXPENDITURE				
	£M	£M	£M	£M	£M
1	)				
2	) 3 over				
3	)				
4	) 5 years				
5	)	3			
6		5	1		
7		10	2	1	
8		15	2	2	1
9			3	4	1
10			3	6	1
11 to 30			(3 pa)	(8 pa)	(1 pa)
(x 19)			57	152	19
	3	33	68	165	22

To present the values over time in terms of their present day equivalent values (NPV – net present value) a simple logic is employed. One pound today is worth more than one pound tomorrow, viz if £1 is invested today it would earn interest and be worth more than £1 at the end of a period of time. Conversely, £1 at the end of a period would be worth less than £1 at the beginning of the period. Discounting is the operation of reducing the values occurring at different dates to a common denominator but immediately, a major problem occurs, it is necessary to consider what would be an appropriate discount rate.

The public sector currently employs a rate of 10 per cent although the private sector may argue a wide range of rates. One person's strength on the money markets will not match precisely the strength of another. For our present purpose it is assumed that a 'perfect market' situation exists, that interest rates will remain constant over time and that price levels will remain unchanged – an assumption that clearly needs close examination in the light of present day economics.

Future price levels could be taken into account by including some arbitrary factor although there would be some difficulty in deciding what that factor should be. However in an expanding or inflating economy costs and incomes are more likely to be affected by a similar magnitude and it is usually safe enough to calculate all values on the present day price levels. It is only where it is felt that some values will move significantly faster than others that the relative differences must be brought out in the calculations. If, for example, oil was expected to continue to rise out of step with other fuels then the difference should be applied in any appraisal. These assumptions allow the whole range of factors to be taken under a single parameter, the rate of interest, which may be considered to contain three components – risk, profit and inflation.

This simplistic approach to an otherwise complex problem is often criticised for being unrealistic. However it is often useful to work on a simple model and then examine how the conclusions of the model may be affected by changing influences. It is usually sensible to carry out an appraisal using a range of interest rates and inflation factors to effect what is known as a sensitivity analysis. If a range of possibilities are examined the sensitivity analysis will show whether significant changes take place that need critical consideration.

The technique of discounting is based on the principle of compound interest utilising the formula

$$\text{NPV} = \frac{\text{Value}}{(1 + i)^n}$$

for a single future value.

If future values occur annually and are of equal magnitude then the formula becomes

$$NPV = \frac{\text{Annual Value}}{(1 + i)^1} + \frac{\text{Annual Value}}{(1 + i)^2} + \text{etc, etc}$$

For convenience there are ample published tables available which avoid the need to calculate the values each time from the formula.

Assuming an interest rate of 10 per cent and returning to the values in the earlier example and applying the appropriate multipliers from the tables the results are as follows:

	Year	*Value £m*	*Discounting Factor*	*Discounted Value £M*
			Assume at end of year 3	
Research	1-5	3	0.7513	2.25
Capital	5	3	0.6209	1.86
	6	5	0.5645	2.82
	7	10	0.5132	5.13
	8	15	0.4665	7.00
		33		16.81
Operating	6	1	0.5645	0.56
	7	2	0.5132	1.03
	8	2	0.4665	0.93
	9	3	0.4241	1.27
	10	3	0.3855	1.16
	11-30	3 pa	9.4270	
			- 6.1450	
			3.2820	9.85
		68		14.80
Income	7	1	0.5132	0.51
	8	2	0.4665	0.93
	9	4	0.4241	1.70
	10	6	0.3855	2.31
	11-30	8 pa	9.4270	
			- 6.1450	
			3.2820	26.26
		165		31.71
Benefits	8-30	1	9.4270	
			- 4.8680	
			4.5590	4.56
		22		4.56

Summary

		Net Value £M	*Discounted Value £M*
1.	Research	3	2.25
2.	Capital	33	16.81
3.	Operating	68	14.80
		104	33.86
4.	Income	165	31.71
		+ 61	- 2.15
5.	Benefits	22	4.56
		83	2.41

NB As. expenditure (items 1, 2 and 3), income and benefits discounted at 10 per cent are almost in balance this indicates a Discounted Cash Flow or Internal Rate of Return of approximately 10 per cent.

With an adverse balance of income on discounted values of £2.15 million the private sector may consider that, as the 10 per cent criterion is not met, the project is not viable as a worthwhile venture. The public sector on the other hand could take the opposite view with a credit balance of £2.41 million after considering the community benefits.

It is possible on this sort of evidence to hold the view that it may be worth public sector support of up to £4.56 million to enable the project to be developed by the private sector – a relevant factor in the appraisal of any ambient energy project. A sensitivity analysis will highlight the effect of changes in the estimated values and bring to the fore the question – how much or what degree of confidence can be attached to the values?

Probability Analysis

Confidence limits can be established by utilising simple probability theory. If one of a number of alternatives is bound to result the alternatives are said to be collectively exhaustive and their individual probabilities will add up to 1.0. If it were impossible for any of the alternatives to result then the probability would be zero.

Applying this theory to the earlier example the first step is to assign to a range of possible alternative estimates a weighting factor representing the probability of their occurrence. This provides a means of obtaining a weighted average or expected value of possible outcomes.

It is more usual to apply probability factors to the values before discounting as the factors may vary from year to year, and total annual figures are valuable data for decision making. However for ease of example the factors are here applied to the discounted values.

1. Research. If it is considered possible that the amount of £3 million over a 5 year period could vary by ± 20 per cent this could be taken on discounted value by the following

NPV £2.25 million	*Percentage Variation*	*Adjusted Value £M*	*Probability Factor*	*Equivalent NPV £M*
	20	1.80	0.05	0.09
	10	2.03	0.15	0.30
	0	2.25	0.50	1.13
	10	2.48	0.20	0.50
	20	2.70	0.10	0.27
			1.00	
			Equivalent NPV	2.29

It has been judged that there is a 50 per cent chance that the original estimate will be proved correct but that there is a 15 per cent and a 5 per cent chance that the estimate will be too high by 10 per cent and 20 per cent respectively. Similarly that there is a 20 per cent and a 10 per cent chance that the estimate will be too low by 10 per cent and 20 per cent respectively. Similar assumptions are made with the other items included in the appraisal.

2. Capital. Variable + 50 per cent to – 20 per cent.

NPV £16.81 million	*Percentage Variation*	*Adjusted Value £M*	*Probability Factor*	*Equivalent NPV £M*
	20	13.45	0.05	0.67
	10	15.13	0.15	2.27
	0	16.81	0.35	5.88
	10	18.49	0.20	3.70

Percentage Variation	*Adjusted Value £M*	*Probability Factor*	*Equivalent NPV £M*
20	20.17	0.15	3.03
30	21.85	0.05	1.09
40	23.53	0.03	0.71
50	25.22	0.02	0.05
		1.00	
		Equivalent NPV	17.85

3. Operating. Variable ± 15 per cent.

NPV £14.80 million

Percentage Variation	*Adjusted Value £M*	*Probability Factor*	*Equivalent NPV £M*
15	12.58	0.20	2.52
0	14.80	0.50	7.40
15	17.02	0.30	5.11
		1.00	
		Equivalent NPV	15.03

4. Income. It is almost certain that income will present a more complex problem. It may be possible to foresee an abnormal increase in energy prices after say the tenth year while being more confident in the price levels assumed for the earlier years. Say for the purpose of the example

Years 7 - 10 possible variation ± 10 per cent

Years 11 - 30 possible variation + 30 per cent
− 10 per cent

	Percentage Variation	*Adjusted Value £M*	*Probability Factor*	*Equivalent NPV £M*
Years 7 - 10 NPV £5.45 million				
	10	4.91	0.15	0.74
	0	5.45	0.60	3.27
	10	6.00	0.25	1.50
			1.00	
			Equivalent NPV	5.51
Years 11 - 30				
NPV £26.26 million	10	23.63	0.15	3.54
	0	26.26	0.35	9.19
	10	28.89	0.25	7.22
	20	31.51	0.15	4.73
	30	34.14	0.10	3.41
			1.00	
			Equivalent NPV	28.09
			Sub-totals	
			Years 7 - 10	5.51
			11 - 30	28.09
			Equivalent NPV	33.60

5. Benefits. Considered constant with inherent errors self-cancelling.

It can prove a useful indicator of overall sensitivity if in an appraisal the range of values is extended by a standard deviation from the equivalent NPV and the results tabulated in summary form, assuming a project period of 30 years.

	Base Estimate		*Sensitivity Analysis*		
	Total	*NPV @ 10%*	*Probability Mean*	*Minimum*	*Maximum*
	£M	*£M*	*£M*	*£M*	*£M*
Research	3	2.25	2.29	2.08	2.50
Capital	33	16.81	17.85	15.37	20.33
Operating	68	14.80	15.03	13.48	16.58
	104	33.86	35.17	30.93	39.41
Income	165	31.71	33.60	30.14	37.06
	+ 61	- 2.15	- 1.57	- 0.79	- 2.35
Benefits	22	4.56	4.56	4.56	4.56
	83	+ 2.41	+ 2.99	+ 3.77	+ 2.21

Minimum/Maximum Range		Minimum	Maximum
	Expenditure	30.93	39.41
	Income	37.06	30.14
		+ 6.13	- 9.27
	Benefits	4.56	4.56
		+ 10.69	- 4.71

The tabulated figures indicate that although the rate of return to the private sector is unlikely to meet the 10 per cent criteria the return to the public sector taking into account the benefits would almost certainly be in excess of 10 per cent and there is little evidence of the risk of failure.

CONCLUSIONS

Application of these techniques to ambient energy projects presents few problems and should assist the understanding of proposals and reports by the expert and inexperienced alike.

There are many developments possible based on these techniques, pricing policies can be established, cost per production unit of energy can readily be compared with any other energy form and by application of consumer efficiency factors the possibly more valid comparison of cost per useful unit can be affected.

These techniques can also be applied to individual energy development projects such as proposals to develop solar energy for domestic buildings and it is perhaps in this field where most people will be familiar with the problems to be considered.

Too many reports fail to give a total cost appraisal. However it was interesting to read in 'Building' 19th November, 1976 a report which included the following:

> "Based on 1976 energy costs and discounting capital costs over a 25 year period the system costs (for a solar energy system providing 40 per cent of the heating requirement and 65 per cent of the hot water requirement) fall between more expensive all-electric and the cheaper combination of electricity and oil-fired space heating. In the long term assuming fuel price rises the conventional combination methods may not be as competitive as solar heated systems".

This report indicates an awareness and probable use of the techniques examined above.

DISCUSSION

Dr. D. Fitzgerald (University of Leeds)

Mr Butler's approach to uncertainty is particularly interesting, but it ought also to be applied to research. One can never be certain of the outcome of research. The possibility of it leading to no benefit at all must be balanced against the possibility of it leading to benefits many times its cost. In the case of the example given, research having a discounted value of £2.25 M with an estimated benefit of discounted value £2.41 M would not have been started.

Mr Butler has used the Treasury Test Discount Rate of 10%. What is the basis of this figure?

What is the average rate of return in British industry? This rate should be the test discount rate.

D. Butler

The example given is a hypothetical one to demonstrate the techniques, chosen to avoid a large margin one way or the other and to illustrate the different approaches there may be between the private and public sectors. The example is artificial to that extent but demonstrates the point about subsidies for example.

The 10 per cent test discount rate is a constant bone of contention. It is a central government figure, the result of lengthy working party discussions. A revised test discount rate, lower than 10 per cent, for public sector use is proposed in a report but the report is only in draft form. There are very strong arguments, in the private sector particularly, for using a lower figure, and some would argue that the figure should be as low as 1½ to 3 per cent when factors such as the return required on investment and the effects of taxation are taken into account. There are arguments, though, for using even a higher figure. For private sector appraisals, the use of the rate of return expected for the particular form of business is to be anticipated. In the public sector, the test discount rate is based on consideration of a complex series of issues and is a central decision. The report referred to is a sizeable document containing full discussion of the various factors involved, but I agree that it is open to argument whether 10 per cent, or any other percentage, is right or not. The figure used hinges largely on a policy decision on whether to spend capital now or money later in running costs, the revenue consequences etc.

G. Foley (Architectural Association)

If 10 per cent discount rate is allowed and inflation runs at 10 per cent, what effect does this have on the evaluation of benefits? It seems that with a 10 per cent inflation, the discount rate should be zero. With 15 per cent inflation, should a negative discount rate be applied?

D. Butler

I will give an illustration to answer this question. If money is not spent on a project, that money is available to spend elsewhere. In business, you invest in that business. A private individual may invest in shares, which is effectively investment in business. If business investment is equated to investment in shares, two returns are expected, a capital appreciation of your shares and a dividend. Take capital appreciation first. Since the late 1800s the movement of the FT Index has, over this long period of time, almost matched the rise in building costs. Perhaps that is not surprising although in the last few years we have been through a period when the figures have moved apart appreciably. In 1972, 1973 and 1974, things like factor costs moved up at a much higher rate than true tender values. Tenders have remained stable for the last couple of years, which is a little surprising when there has been considerable inflation in other fields. Taking the period from 1890 to 1970, the FT Index and the relative rise in building costs were on a par except during very short periods. In this work, we are concerned with the long term. The factor of capital appreciation accounts for inflation. The dividend return expected on shares is the factor I think we are using in our discount rate. A differing percentage is expected at differing times and this in part is why the test discount rate will vary. I see the two returns and the one ignored in calculations is the capital appreciation, which offsets inflation. Another factor, of course, is that figures are all relative and so prices in the future do not matter quite as much as might be anticipated. The only time that this aspect needs to be considered, and I have made a point of this, is when something like oil changes price at a rate unrelated to inflation and the general movement of the prices of other commodities. If something like that can be foreseen, it must be taken into account. It certainly needs to be taken into account with fuel because as a result of shortage it will become more expensive at a greater rate than other things over a period of time.

Dr. J. C. McVeigh (Brighton Polytechnic)

An important point is made by drawing attention to the failures of economists in the past and particularly to the 1967 White Paper on Energy Policy. It is clearly very difficult to quantify the future value of oil and gas as the reserves become depleted. Mr Butler emphasised the need to be aware of complacency, but it is by no means certain that supplies of North Sea oil and gas will necessarily last until the year 2000. It depends entirely on demand.

Perhaps the greatest economic myth of our time is the view that the gross national product must be increased and that to do this energy consumption must be increased. This is nonsense on both counts. I believe we can improve our quality of life, by moving to a decentralised society as outlined, for example, in the book "Small is Beautiful" by Schumacher and in "Man, Machines and Tomorrow" by Thring.

The question of quantifying a benefit is an interesting one. How large a negative value would be placed against nuclear power in twenty years time by those whose children were suffering from the effects of exposure to radioactive material? I believe it is possible to have a society that depends largely on renewable energy resources. I am hopeful that research will confirm this view and the political backing that is necessary to develop these renewable resources will be available.

4. Usable Energy from Wind

D. F. Warne

INTRODUCTION

Wind power can be used for a wide variety of purposes, from pumping water to heating greenhouses, charging yacht batteries, supplying remote telecommunications repeaters and, at the megawatt end of the scale, delivering directly to a high voltage electrical grid. A paper on usable energy from the wind could therefore take many forms; the intention here is to provide some guidance on the selection of small and medium size wind generators that are commercially available, with some indication as to the extent of their applicability.

Historically, small wind generators were first successfully marketed in significant numbers around the turn of the century, when pioneer farming and other 'outback' activities had a need for electricity ahead of or in the absence of the rural spread of utilities. This market was most notable in the USA and Australia, where a proliferation of small plant became commercially available. The early types incorporated turbines with a large number of flat metal blades; the first world war encouraged a rapid development of aerodynamics theory and technology and most manufacturers subsequently turned to the low solidity propeller – like turbine for its increased specific power output and reduced material requirement. (Water pumping turbines, with some exceptions, retain the older style largely for its superior starting torque).

The market for the small wind generators (typically up to 5 m turbine diameter) collapsed fairly rapidly in the late 1940's with the growth of electrical supply networks and during the 1950's and 1960's only a few manufacturers remained in business serving specialised markets.

The fuel shortages of the post-war period motivated several national wind power development programmes, but these were mainly aimed at large scale grid supply. Several manufacturers were tempted in the course of this work to enter production on small wind generators, but few achieved any measure of success. Notable exceptions perhaps were German machines marketed under the 'Allgaier' trade name; these machines (up to 10 m turbine diameter) were technologically sophisticated but gave mixed operational experience, about 100 being produced. (Fig 1 shows a 10 m diameter unit that the Electrical Research Association (ERA) tested in the late 1950's as part of an autonomous domestic power supply).

The energy crisis of 1973 provoked fresh interest in wind power throughout the range of power and applications, and several new manufacturers have entered the field alongside the established firms, some of whom have now been in business for nearly fifty years. There are consequently a large and still growing number of machines available, differing widely in their characteristics, quotable operational experience and applicability in different circumstances.

BASIS FOR ASSESSING WIND GENERATORS AGAINST INDIVIDUAL APPLICATIONS

The simple generalised momentum theory for axial flow turbines indicates that the power output W from the turbine is given by

Fig 1 **7.5 kW Allgaier machine installed by ERA to provide autonomous power to a house at Achnagoichan, Aviemore.**

$$W = \frac{1}{2} C_P \rho A V^3$$

where ρ is air density, A is the frontal swept area of the turbine and V is the free wind speed well upstream of the turbine (SI units). The constant of proportionality C_P is known as the power coefficient, and the simple momentum theory proposes a maximum value of 16/27. Experience has indicated that values of C_P up to 0.46 - 0.47 can be achieved in large highly efficient turbines, but smaller turbines give generally lower values. Small machines suffer again because of the relatively poor efficiency of small gearboxes and generators, and when these losses are accounted for, an overall output power coefficient C_{OP} of between 0.2 and 0.3 is common in machines up to 5 m to 10 m turbine diameter.

Assuming C_{OP} is 0.25, and air density is 1.28 kg/m^3 (this varies marginally according to temperature and pressure) Table 1 indicates the power output that can be expected from plant having turbine diameters of 1 m, 5 m and 10 m at a range of wind speeds.

Table 1 **Influence of wind speed on output of turbines of various diameters.**

Turbine diameter (m)	*Electrical power output (W)*		
	Wind speed 3 m/s	*Wind speed 7 m/s*	*Wind speed 12 m/s*
1	3	43	217
5	85	1080	5420
10	340	4310	21700

The arithmetic behind Table 1 is trivial, but the output figures serve to illustrate the overwhelming influence of wind speed. The wind speeds have been selected because they represent the annual mean wind speeds available in the UK:

3 m/s – fairly commonly over inland areas
7 m/s – only with careful siting and usually near the coast
12 m/s – only at a few of the windiest coastal hilltop sites

The output figures speak largely for themselves, but it is worth pointing out that a 10 m turbine operating in a 3 m/s wind may produce only one third the output of a 5 m turbine on a 7 m/s wind. Local topography can have serious effects on wind speeds, and the installation of a wind generator should be preceded by careful study of the sites available, preferrably assisted by long term anemometry.

The output power of a wind generator, unlike that of a diesel or steam turbine generator, is to a large extent uncontrollable and varies with instantaneous wind speed. Most manufacturers issue a power curve indicating the relationship of power output to wind speed. Fig 2 shows a typical curve. Since the output power cannot be relied upon at any instant, the important performance figure is the cumulative energy output over a period long enough to be representative of the frequency of wind speeds at a chosen site. This period is usually taken as one year in order to cover seasonal variations, although longer periods will give greater certainty of course.

The appropriate representation of the wind speed distribution for purposes of energy prediction is known as a velocity duration curve (Fig 3) which indicates the number of hours pa for which wind speed exceeds various levels. For a particular plant at a particular site, each ordinate of the velocity duration curve can be translated using the plant power curve (Fig 2) into an ordinate of the power duration curve (Fig 4). The shape of the power duration curve varies substantially between plant-site combinations, depending upon the relationship of plant cut-in and rated wind speeds to the appropriate velocity duration curve. The important factor is the area beneath the power duration curve, which gives the annual energy output.

It will be seen in the following section that wind generators as currently available have a range of cut-in, rated and furling wind speeds, which make some more suitable for any given site than others. At one extreme, if a machine with high cut-in and rated wind speeds is selected for a low mean wind speed site, it will have a very low utilisation factor and will produce little annual energy (this corresponds to a power duration curve 'shifted' well to the left). At the other extreme, if a machine with low cut-in and rated wind speeds is selected for a high mean wind speed site, it will regulate its output and 'spill' power for a large proportion of the time, producing again less annual energy than is practically possible (this corresponds to a power duration curve with a long, but low, horizontal section).

ERA has evolved the concept of an Energy Coefficient C_E to assist in assessing the quality of 'matching' of a wind driven unit to a given site. The method of calculation is as follows.

In addition to the power duration curve constructed for the practical wind generator, using its own characteristic and the velocity duration curve for the site, a second power

Fig 2 Power-wind speed characteristic for typical plant.

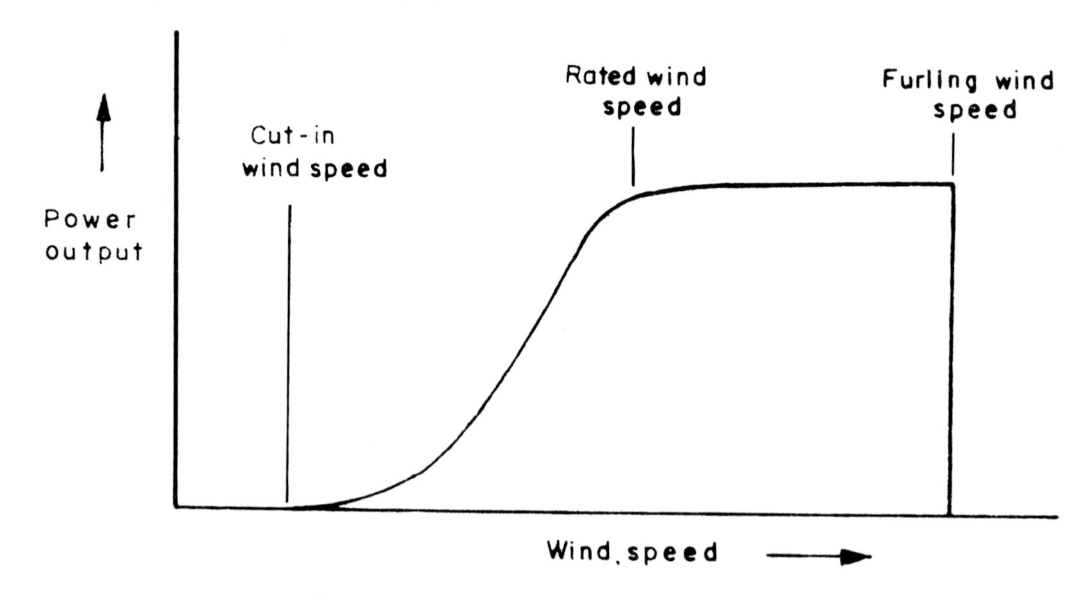

Fig 3 Typical profile of a velocity duration curve.

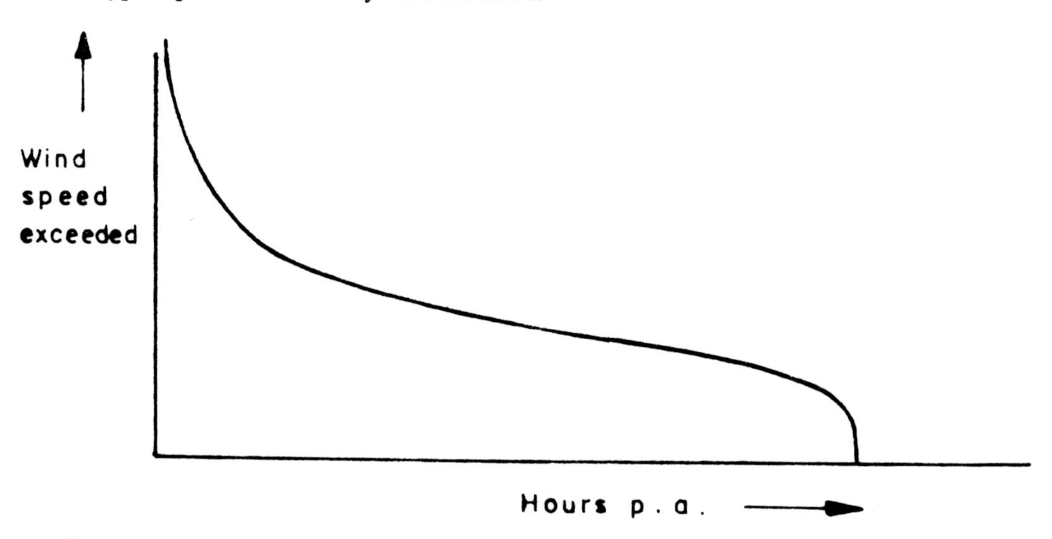

Fig 4 Power duration curve.

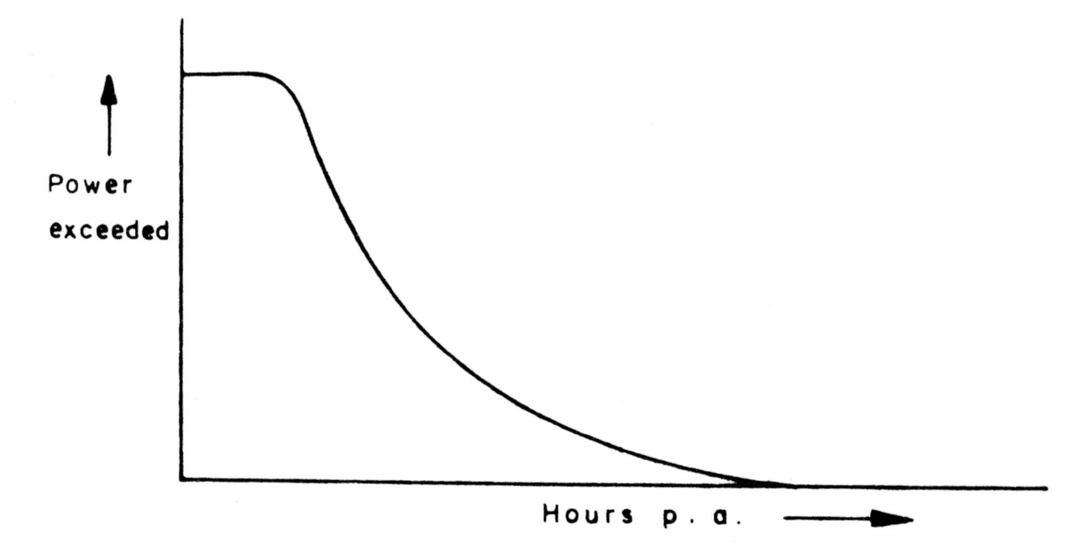

duration curve is drawn. Using the same velocity duration curve, the ordinates are translated into power ordinates on this second curve assuming a turbine of equal area A to the practical unit, but having unity power coefficient C_P so that

$$W = \tfrac{1}{2} \rho A V^3$$

The second curve is in a sense the 'ideal' power duration curve for the particular velocity duration curve and turbine area selected.

The energy coefficient is defined as the ratio of the annual energy outputs calculated from the area under the two curves,

$$\text{ie } C_E = \frac{\text{annual energy output from practical plant}}{\text{annual energy output from plant having equal A but } C_P = 1}$$

It has been ERA's experience that optimum matching (ie maximum value of C_E for any practical plant) occurs when the site annual mean wind speed is between 50% and 70% of the plant rated wind speed. For good matching, C_E should be around 0.2 or above for the large machines (5 to 10 m turbine diameter) and slightly lower for smaller machines. (ERA has found that bad matching can result in C_E values as low as 0.05, even for large machines).

PRACTICAL WIND GENERATOR REVIEW

Table 2 compares the products of several different manufacturers of wind generators. In this context it is obviously impossible to provide a comprehensive list of all such products, and those mentioned here have been selected for illustrative purposes only. They include some items from manufacturers who have a protracted history in the field and whose products are the result of extensive research, development and practical experience, and other items from manufacturers whose products have been introduced more recently.

The manufacturers considered in the table have also been chosen from a variety of countries so as to provide a truly international picture. It must be emphasised that the inclusion of a manufacturer does not imply that his products are superior to those of other manufacturers who are not mentioned.

The table picks out the prime performance and technical data associated with the plant in each manufacturer's range. Data are often not obtainable directly and some have been interpreted from information supplied by the manufacturers. (Plant at present under development is excluded from Table 2.)

Equally important to these technical data are costs and histories of operational experience. The current economic situation makes a definitive listing of costs inappropriate, particularly since most of the units are imported. (Economics are commented upon in the final section, but the approach is to reverse the analysis and consider the breakeven capital expenditure that can be justified). Operational experience is difficult to catalogue definitively, even with ERA's extensive network of contacts throughout the world. In assessing reports of failures, several moderating factors must be considered, such as the objectiveness of the report, the likely engineering skill of the user in specifying and maintaining the plant, and the skill of the construction and commissioning personnel. It is also worth remembering that the older established firms have sold several thousand machines, and the occasional failure may not be statistically significant (reports on 20 years successful operation are seldom written, although such experience is known).

ENERGY STORAGE

Although the energy output from a wind generator over long periods can be predicted with some certainty in the manner described above, the short term power output cannot

Table 2 **Principal data associated with some commercially available wind generators.**

Manufacturer	*Model*	*Rated power (W)*	*Rated wind speed (m/s)*	*Cut-in wind speed (m/s)*	*Turbine dia (m)*	*No of blades*
Aerowatt SA	24FP7	24	7	4	1.2	2
37 rue Chanzy	100FP5	100	5	2	3.2	2
75011 Paris	150FP7D	130	7	3	2.0	2
France	300FP7B	350	7	3	3.2	2
	1100FP5	1125	5	2	9.2	2
	1100FP7B	1125	7	3	5.0	2
	4100FP7	4100	7	3	9.2	2
Dominion Aluminum Fabricating Ltd 3570 Hawkestone Road	'15 ft'	4000	12	3	4.5	2
Mississauga Ontario Canada	'20 ft'	6000	12	3	6.0	2
Dunlite Division of Pye Industries PO Box 120 Oakleigh Melbourne 3166 Australia		2000/ 3000	11	4	3.2/ 4.0	3
Elektro GmbH	WVG05	600	8	3	2.5	2
8400 Winterthur	WVG15	1200	10	3	3.0	2
St Gallerstrasse 27	WVG25	2200	12	3	3.6	2
Switzerland	WVG35	4000	11	3	4.4	3
	WVG50	6000	13	3	5.0	3
	WVG120	10000	15	3	6.8	3
Enag SA	1	650	9	–	–	2
Route de Pont L'Abbé	2	1200	9	–	–	2
29000 Quimper France	3	2500	9	–	–	2
Lübing Maschinenfabrik 2847 Barnstorf Bez Bremen Postfach 110 West Germany	M-022-3 G-024-400	400	10	4	–	3
Quirks Victory Light Co (Pty) Ltd 33 Fairweather Street Bellevue Hill NSW 2023 Australia		3000	11	3	–	3
Wind Energy Supply Co Ltd Bolney Avenue Peacehaven Sussex BN9 8HQ		5000	14	4	5	2
Winco Division of Dyna Technology Inc PO Box 3263 Sioux City Iowa 51102 USA	1222H	200	10	3	1.8	2

be relied upon. If the wind generator is to be part of a firm power supply, it is therefore essential that it be coupled to some form of energy storage facility.

The form of this energy storage may vary according to local circumstances and the final form of power output required (principally, whether this is electrical or thermal). The hardware problems of energy storage are not peculiar to wind power systems, and will not be elaborated upon here, either in their thermal or electrical form. Attention is focussed on the more fundamental problem of determining the capacity of storage which will be required in a specific application.

A simple and traditional technique associated with wind power systems has been to allow for a five day period of zero wind generator output. This assumes that prior to the five day period the wind generator has been producing enough power to supply the load and fully charge the storage system, and that following the five day period the wind generator will immediately produce enough power to completely meet load demand; it does not resemble a likely occurrence in practice. Different combinations of wind patterns and load demand profiles, coupled with a variety of possible wind generator choices to meet the same overall energy demand will result in very different requirements for storage capacity. The five day rule, if it results in practice in inadequate capacity is an engineering failure. If it results in over-capacity it is a technoeconomic failure and the initial capital expenditure is higher than necessary.

Recognising these shortcomings of the simple technique, ERA has developed a method for determining the required storage capacity given a particular site, wind generator and load demand profile. The method is best described with reference to Fig 5. A representative real time pattern of wind speed over a protracted period is first established, and a corresponding real time power output variation from the wind generator calculated. Making any due correlation with wind speed, a load profile is proposed for the same time period. A net power output profile can then be calculated, and this is then integrated with respect to time to produce a net energy surplus/deficit profile. Storage capacity is determined by the peak surplus or deficit level, and making due allowance for a minimum state of charge that may be tolerable (particularly for electrolytic batteries).

In principle this technique is an ideal analytical tool, but there are several practical problems. Two notable difficulties are the establishment of a truly representative wind pattern, and fixing a meaningful datum level on the energy surplus/deficit profile. The former is a statistical problem, and with sufficient wind data can be satisfactorily resolved. The latter problem relates to the 'start point' of the system; the net energy balance curve is not usually as exaggerated as that shown in Fig 5, but this problem does nevertheless deserve serious consideration. It scarcely need be mentioned of course that an accurate definition of the load demand profile is necessary. Without this no precise estimation of storage requirements is possible.

Since the analysis method is most amenable to computer-based calculation, it is usually convenient to test the sensitivity of the answer to extreme differences in the pattern of wind speed selected; this can be used as a guide to the safety factor incorporated in the ultimate choice of storage capacity.

As a final comment on storage of energy, it is worth commenting that in certain applications a fuel-based power source (such as a diesel generator) can be used to provide reserve power as complete or partial replacement for a passive storage system. The design procedure then becomes more complex, involving a series of technical and economic trade-offs between the proportions of fuel-based, storage and wind generator capacity necessary to meet the overall energy demand.

ECONOMICS

A review of capital costs would clearly date very quickly in the current economic situation, and in any case the price of the wind generator itself is only part of the total instal-

Fig 5 Real time wind-speed, power and energy pattern analysis.

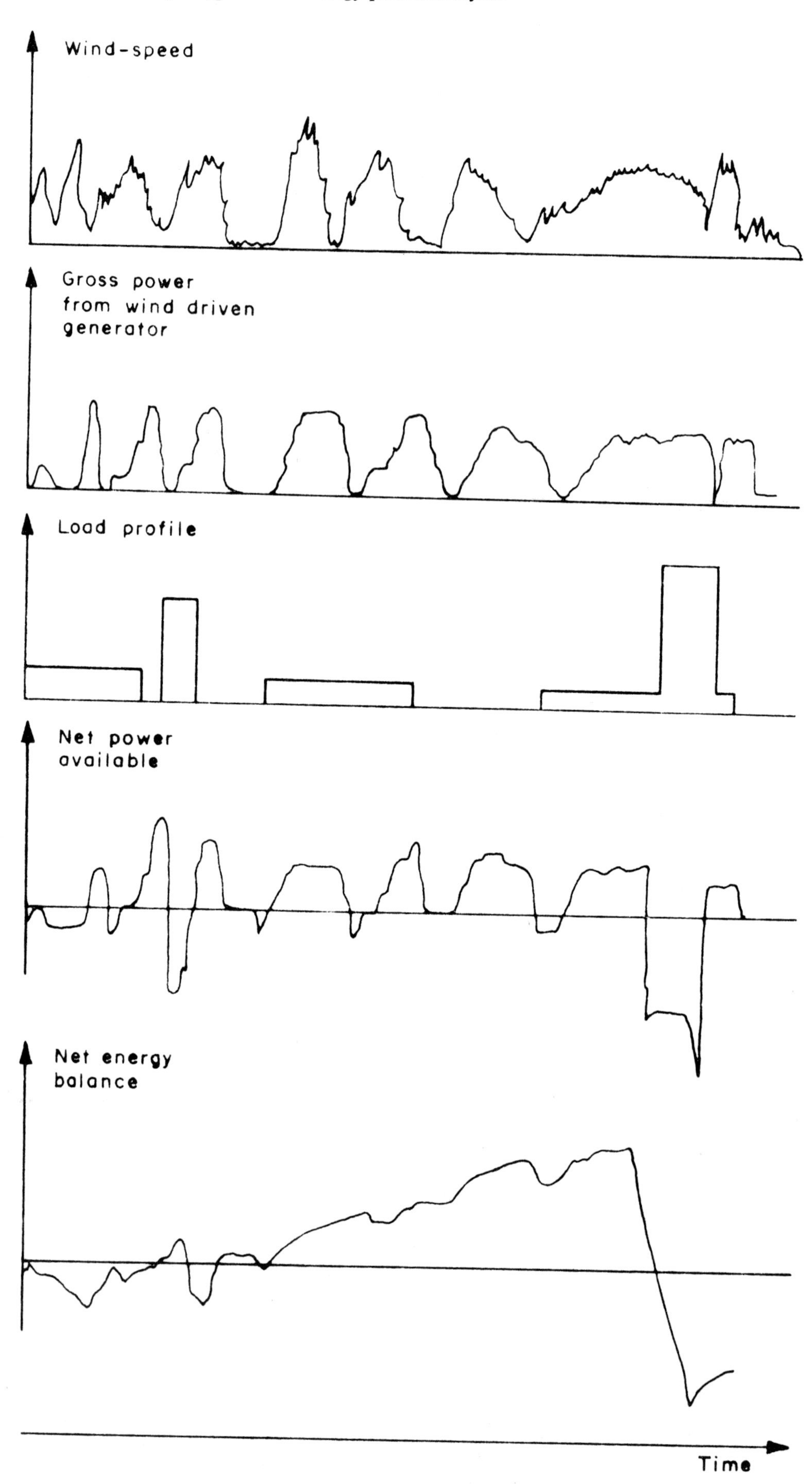

lation cost; this fraction may sometimes be quite small if the location is difficult or special system features are required. Without discussing specific application examples, an economic judgement may more simply be made by turning the approach around. That is, to take a particular type of wind generator, to attach a value to its annual energy output for several mean wind speeds, and using an acceptable capital recovery period to determine the capital cost that could be justified.

As an example, the Elektro model WVG 50 will be considered. Using a normalised velocity duration curve which is reasonably typical in shape, and factoring this up for different annual mean wind speeds, the annual energy output of this machine is predicted in Table 3.

Table 3

Annual energy output of Elektro model WVG 50 at various annual mean wind speeds.

Annual mean wind speed (m/s)	*Annual energy output (kWh)*	*Electrical energy value pa at 2p/kWh (£)*	*Justifiable capital (5 year recovery period, 10% annual charge) (£)*
3	2200	44	167
5	9000	180	683
7	17900	358	1357
12	30600	612	2320

The influence of mean wind speed is again seen to be critical. 5m/s has been added to the three wind speeds of Table 1 because of the particularly wide and signficant gap between the energy output figures at 3m/s and 7m/s mean wind speeds.

Valuing the output energy at 2p/kWh, the approximate annual return figures are also shown in Table 3. Several accounting methods are of course available to calculate the justifiable capital expenditure from these annual returns, given an acceptable capital recovery period. A five year capital recovery term and a 10% effective annual interest charge have been assumed for the purposes of illustration. (This chapter is not the appropriate place for discussion of either a valid annual charge to account for inflation and maintenance costs or what is and what is not an acceptable capital recovery period, or even whether energy return should be considered instead of financial capital return). The resulting justifiable capital costs are shown in the final column of Table 3.

The January 1977 price of an Elektro WVG 50 (through UK agents for Elektro, Conservation Tools and Technology) was approximately £4000. The purpose of the simple calculations on justifiable capital outlay is to highlight the difficulty of justifying the installation of a unit against grid-produced electricity in a simple 'fuel-saving' mode and at current wind generator prices, particularly when the additional costs of installation, storage and system interface are considered.

Although prices of other units are different (and the calculations for thermal output machines must be based on different energy values) the broad conclusion is nevertheless that on a hard accounting basis it will usually be difficult to justify a wind generator against grid-produced electricity. This is not to say that many viable applications do not exist, particularly for remote autonomous power supplies where effective alternate energy costs are often extremely high, or where a substantial charge for electricity connection can be offset against the capital for the wind generator, or where the user may be concerned with energy independence regardless of cost.

It will be important to continuously up-date the economic appraisal of small wind generators against future changing balances of energy and material costs; it is hoped that this paper will provide assistance on such future evaluations of specific cases.

ACKNOWLEDGEMENTS

I am grateful to the Directors of ERA for permission to publish this chapter and to colleagues at ERA, particularly Mr P G Calnan, for their assistance in compiling the information on which the paper is based. Especially I am indebted to the late Mr A H Stodhart for the benefit of his extensive experience in this field.

DISCUSSION

Professor J. K. Page (University of Sheffield)

We are working with the Building Research Establishment on the measurement of wind on a nineteen storey building in Sheffield. I have some preliminary data for three levels and I find the energy amplification factor at the top of the building is 20, when compared with the standard meteorological station height of 10 m. Do the Electrical Research Association think it would be possible to put windmills on buildings?

D. F. Warne

The main problem with siting windmills on buildings is one of turbulence – it is indeed possible to find high acceleration factors at increased altitudes on the top of buildings but it can be accompanied by a destructive amount of turbulence which in many cases it may not be possible to economically build wind turbines to withstand.

Dr. J. C. McVeigh (Brighton Polytechnic)

When considering windmills which will last for 15 years we must evaluate the financial returns over that period and certainly not over much shorter periods such as 5 years.

Mr Warne mentioned that wind power can be used for a variety of purposes. I would like to add some further points.

(i) The match between the availability of wind power and seasonal demand. This is particularly true for heating glass-houses.

(ii) Because of the moderate economy of scale, wind power units will be relatively inexpensive if mass produced. A windmill is about one sixth as complex as a motor car. Perhaps it would be worthwhile investigating the social implications of changing from the production of large cars to the building of small windmill units.

(iii) Wind energy is high grade energy. It can be used as mechanical power, electrical power or degraded directly to low temperature heat.

(iv) Wind power can be used with existing systems and does not require changes to facilities or to lifestyles.

(v) No new fundamental technology is needed for horizontal-axis windmills as development can draw very heavily on existing knowledge in allied fields.

(vi) About 80% of a typical modern windmill consists of conventional equipment that is already available. The development costs of the remaining parts will be relatively low.

(vii) Export potential is substantial.

(viii) Of all the renewable energy possibilities mentioned in the document produced by the Advisory Council on Research and Development for Fuel and Power (Advisory Council on Research and Development for Fuel and Power. Energy R & D in the United Kingdom. A Discussion Document. HMSO: 1976), wind power is perhaps the one that provides the most rapid, the least capital intensive and most reliable replacement for imported fuel.

In the outer regions of the United Kingdom, wind power could supply energy for domestic or district heating systems, for agriculture, or for desalination systems, if necessary, at a cost considerably less than some of the Islands of Scotland are having to pay for their energy at this moment. In places such as the Isles of Scilly, the Isle of Man and the Channel Islands, the cost to the consumer of wind-powered energy could be very reasonable, lower than with conventional plant using fossil fuels.

Much is made of the visual impact of windmills, but this depends entirely on the size of windmill and siting. The siting of giant vertical axis windmills in the Wash has been proposed. There the sea is comparatively shallow and not very rough. Horizontal axis windmills with blades of about 46 metres diameter have been designed and the aesthetic challenge can be met by structures similar to the "Skylon", which was a central feature at the Festival of Britain. We have become blind to the visual impact of thousands of electricity pylons across the countryside, from every power station.

D. F. Warne

I would not defend the five year capital recovery period, but both industrialists and domestic users do have investment alternatives. With a shortage of capital, the one that repays most quickly is taken.

R. Cullen (Architects Design Group)

At the Institute of Landscape Architects Conference held in Glasgow in 1975, the Hydro Board and their landscape architects presented a paper about alternative routing of power lines across wide and beautiful landscape in Western Scotland to the Isle of Mull. The present timber poles and 33 kV supply was inadequate and insecure. The Conference, almost to a man, said that it would be criminal to adopt either route and it was asked whether the Board had looked at alternative ways, particularly wind, of generating electricity on the island. The reply from the Hydro Board was that there was no way of guaranteeing a supply using alternative technology.

Is there a way that an Island on the West Coast of Scotland could be provided with this power now?

D. F. Warne

The technology is available. It is a question of economics – what is the energy cost with and without a power line and what is the present cost? The mean wind speed in the Isle of Mull is certainly among the highest available in the United Kingdom, and there is no technological barrier.

Dr. R. W. Todd (National Centre for Alternative Technology)

A system is at present being designed for the Magdalene Islands. The use of 100% windpower is aimed at, pumped water storage from underground caverns being used to give a fairly long holdover time. The scheme could well be cost effective when compared with the cost of diesel power generation, after allowing for the cost of shipping out the diesel oil.

D. F. Warne

The scheme for the Magdalene Islands referred to was the subject of a feasibility study. There is no definite plan to go ahead with the scheme but it has been described in a paper presented at the BHRA Symposium in September, 1976. There is a major project going ahead on Magdalene Island, the erection of a 200 kW vertical axis turbine which will operate in parallel with existing diesel generation capacity.

Dr. G. S. Saluja (Robert Gordon's Institute of Technology)

The problem of storage of electricity generated by a windmill can be neatly solved if the system can be linked with the national supply grid. I would be interested to know if there is any possibility of arrangements being made between the owner of a windmill and the Electricity Board.

D. F. Warne

This depends of course upon the attitude of the local Electricity Board. One of the problems with most small machines is that they are variable speed and variable frequency. The WESCO machine is an exception because it has a hydraulic drive giving the possibility of a fixed frequency output but otherwise one is confronted with using an inverter. The possibility of harmonic and spike generation from inverters may concern local Electrical Boards to the extent that they will demand a low noise inverter, which could be extremely expensive.

A. W. K. MacGregor (Napier College, Edinburgh)

There might be potential for coupling wind power with storage in some of the existing pumped storage schemes in Scotland.

D. F. Warne

There is the potential but I think that when the power system use is examined, the storage capacity is found to be already being used to near its full extent. Its use has to be discounted against variation of load demand and it cannot also be discounted against variation of wind power input.

Dr. R. W. Todd

Wind generated electricity can already be sold to the Utility in several parts of the United States and there is a pilot scheme under way in this country. I believe one Electricity Board is accepting power up to 5 kW into a local 11 kV line. The financial arrangements are tentative at the moment but it looks as if the amount paid for electricity will be roughly the fuel cost, which is about one third of the normal cost to the consumer.

D. F. Warne

I believe that the Electricity Board concerned required either 5 kW or no power, which probably created a design problem and made the system more expensive than need be. I am not sure of the reason for this particular requirement.

5. Active Collection and Use of Solar Energy

B. J. Brinkworth

INTRODUCTION

Of all routes to the utilisation of ambient energy, the direct use of the solar radiation for heating has evoked the most public interest. Here is an energy source which appears to have none of the disadvantages of the conventional fuels or of nuclear energy; it is perpetual, silent and non-polluting. Moreover, it is distributed freely to all premises and there appears to be no way in which it could be interrupted for political reasons or in furtherance of an industrial dispute. Its use would confer a measure of independence, at both national and individual levels, the attractions of which have been universally recognised.

The instinctive enthusiasm which these characteristics arouse makes it all the more necessary that the matter be approached objectively. The many advantages which solar energy has over the conventional sources do not include superior economics at the present time, except at low latitudes (1). Though the balance continues to shift in its favour, the value of the fuel saved by its use has not yet gained a marked ascendancy over the capital and running costs of installations for collecting and using it. Fortunately, there appears to be a growing body of sponsors and users, in all parts of the world, who look beyond the immediate values of the day. Through them, there is a steadily-accelerating programme of system development and accumulation of experience. This should ensure an orderly phasing-in of this new energy source, rather than a hasty scramble when the exhaustion of the older ones becomes imminent. Conferences which encourage a wider understanding of the possibilities and the problems, have an important part to play in helping to smooth the transition into the new energy situation that approaches.

SOLAR ENERGY USE IN BUILDINGS

The characteristics of the solar radiation are the subject of other chapters and need not be elaborated here. It suffices, to explain the worldwide interest in it among those concerned with building design, to note first that the energy delivered by it at sea level over the year exceeds 3500 MJ per square metre of surface for most of the inhabited parts of the earth. This means that the energy intercepted by the external surfaces of a building is of the same order, and in many situations much greater than, that which is used for human purposes inside it. These purposes are principally domestic water heating, space heating and space cooling.

Of these, the use of solar energy for space cooling is perhaps the most fitting, since in that application the greatest demand occurs at the time of greatest energy supply. However, thermodynamic limitations severely restrict the effectiveness with which solar energy can be used for cooling. For systems employing compression-type machines, for example, a typical maximum cooling load that might be sustained would be about 10 MJ/day per m^2 of collecting surface, with heat rejection at 30°C. For absorption-type machines, the corresponding value is about 7 MJ/day per m^2 (2). It follows that very large collection areas would be required, even for domestic use, and with such a seasonal load, the economics are not particularly attractive.

Space heating represents another seasonal load, but one which is out of phase with the solar energy supply. In consequence, a large collection area is required if the system is to provide a significant proportion of the heating requirement in winter, and then the excessive collection in summer may become an embarassment (3). It has long been recognised that inter-seasonal energy storage would greatly enhance the possibilities of solar space heating, but this has so far been hampered by high cost and awkward space requirements. Studies are in hand of ways of minimising these (4) and the prospects for effective long-term energy storage are reviewed elsewhere. There is a growing interest in architectural design which maximises the capacity of the building itself for the collection and storage of solar energy. This is being distinguished from the arts of the building services engineer by calling it passive collection; where a system is installed specifically for collection and use of solar energy, this will be termed active collection.

The most familiar active system is aimed at the heating requirements for domestic hot water, which are fairly constant throughout the year in many countries. Solar energy can be applied to the reduction of conventional fuel consumption in meeting these requirements, carrying all or most of the load in summer, and contributing in some degree at other times. In Britain, domestic hot water heating accounts for some 7% of the national energy consumption (5) and this alone is a worthy target for measures aimed at energy conservation and reduction of living costs.

SOLAR WATER HEATING

Solar water heaters have received a good deal of publicity in recent years and have been very thoroughly described in the literature (6). Yet a good deal of uncertainty remains, concerning their performance in a given environment and the optimum sizing of components for a given duty. Some forty firms are currently offering solar collectors and systems for hot-water heating in Britain, and their literature presents a bewildering variety of performance predictions and advice to the customer. Even the collector area recommended for a typical family dwelling house varies from as little as 3 m^2 to as much as 8 m^2. Performance predictions are often given in terms of the collection efficiency, or the fraction of the incident energy which is retained in the heated water. Nothing can be inferred from this in the absence of detailed information about the conditions under which it is achieved. In particular, it cannot provide a meaningful basis upon which different products may be compared. Very few firms have been able to present performance figures in terms that may be readily understood by the customer, such as a month-by-month tabulation of the conventional energy consumed by the solar-assisted system and by a similar one without this assistance.

Part of the difficulty arises from the newness of the industry and the lack of user experience with systems of this kind. Even where systems are fully monitored, however, the accumulation of data is very slow and does not provide a satisfactory basis for rapid development and innovation. This may come through a combination of monitored performance and the use of computer modelling programmes. Once a programme has been validated by satisfactory comparison of its predictions with the performance of monitored systems, it may be used to determine the optimum combination of component characteristics for a given duty and the results of actual or potential changes in the technology. This is the philosophy being followed within the Solar Energy Unit (7).

A comprehensive range of test facilities is being built up to assist in this process. Outdoor test work has been carried out on a rig built for a collaborative test programme with other laboratories throughout the EEC and this is being supplemented by a more versatile facility being developed under the aegis of the International Energy Agency. Indoor tests are made for the determination of loss coefficients and radiative characteristics, with measurement of the performance of components and systems following in the Unit's solar simulator. Live performance is being monitored in a system installed in a 4-storey block of students' flats. A parallel development of prediction programmes is continuing, and some results arising from this are given later.

System characteristics

The majority of solar water heating systems available in Britain are varieties of the generic system shown in Fig 1. The fluid heated in the collector panels may circulate within a closed circuit containing a heat exchanger in, or sometimes external to the storage tank – in a direct system the heat exchanger is absent and the water itself passes through the collector panels. A separate vessel is shown, in which an auxiliary heat source raises the solar-heated water to tap temperatures, though in some cases this vessel may simply form the upper part of the storage tank. When hot water is drawn from the auxiliary vessel, it is replaced by pre-heated water from the storage tank rather than from the cold supply. When the storage tank is at or above the required tap temperature (say 60-65°C) no auxiliary heating is needed, but for much of the year in a typical system the storage tank temperature will be well below this.

Figure 1 **Typical solar water heating system.**

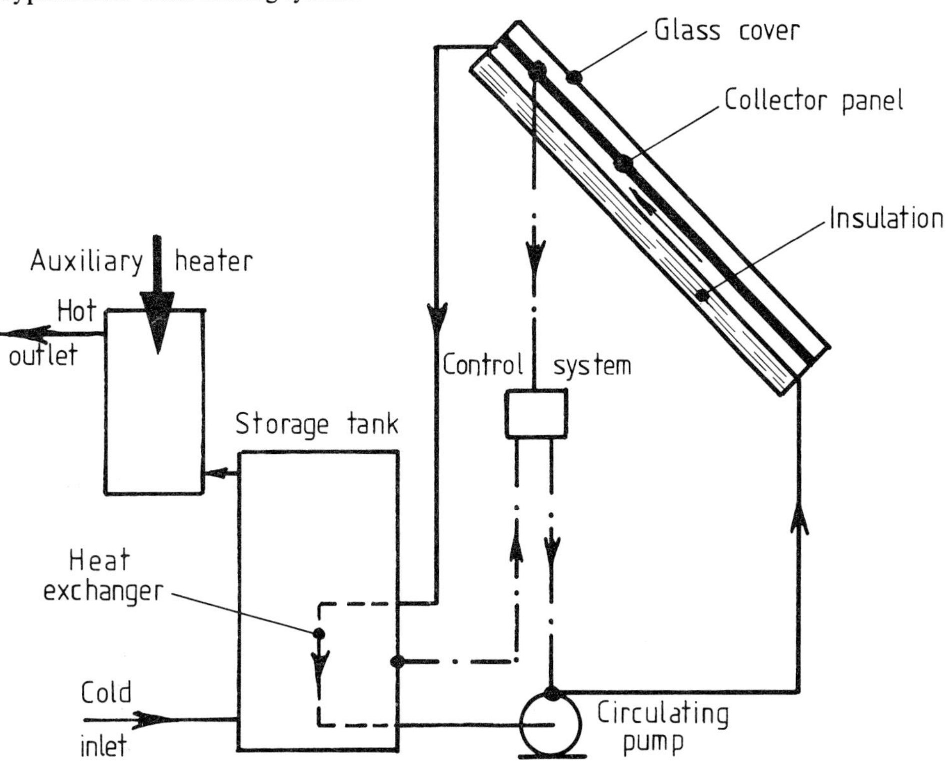

Various arrangements are employed for dealing with potential boiling and freezing, both of which can occur in Britain. In indirect systems boiling may be suppressed by pressurisation or the use of high boiling-point fluids and freezing by anti-freeze additives. Direct systems are usually so arranged that the collector panels are drained when either boiling or freezing is likely to occur. This is governed by the control system, which also ensures that there is no circulation through the panels unless an energy gain would result from this. It is usually arranged that the pump is actuated when the temperature rise at full flow in the panels is greater than some value, selected so as not to cause undue hunting – typically 1 to 3°C.

Much can be visualised about the behaviour of such a system by considering what happens in a collector panel. Generally speaking, it can be said that if the panel temperature (determined mainly by the temperature of the fluid entering it from the storage tank) is above the ambient temperature, energy will be lost from it to the surroundings. Only if the flux of solar energy reaching the panel is greater than this loss can there be a nett gain of energy. A typical relationship between the energy gains and losses is illustrated in Fig 2, which shows the effects of some of the most significant factors such as fluctuation in the solar radiation, the user's demand for hot water and the temperature rise setting of the control system.

Figure 2 **Energy fluxes for solar collector panel.**

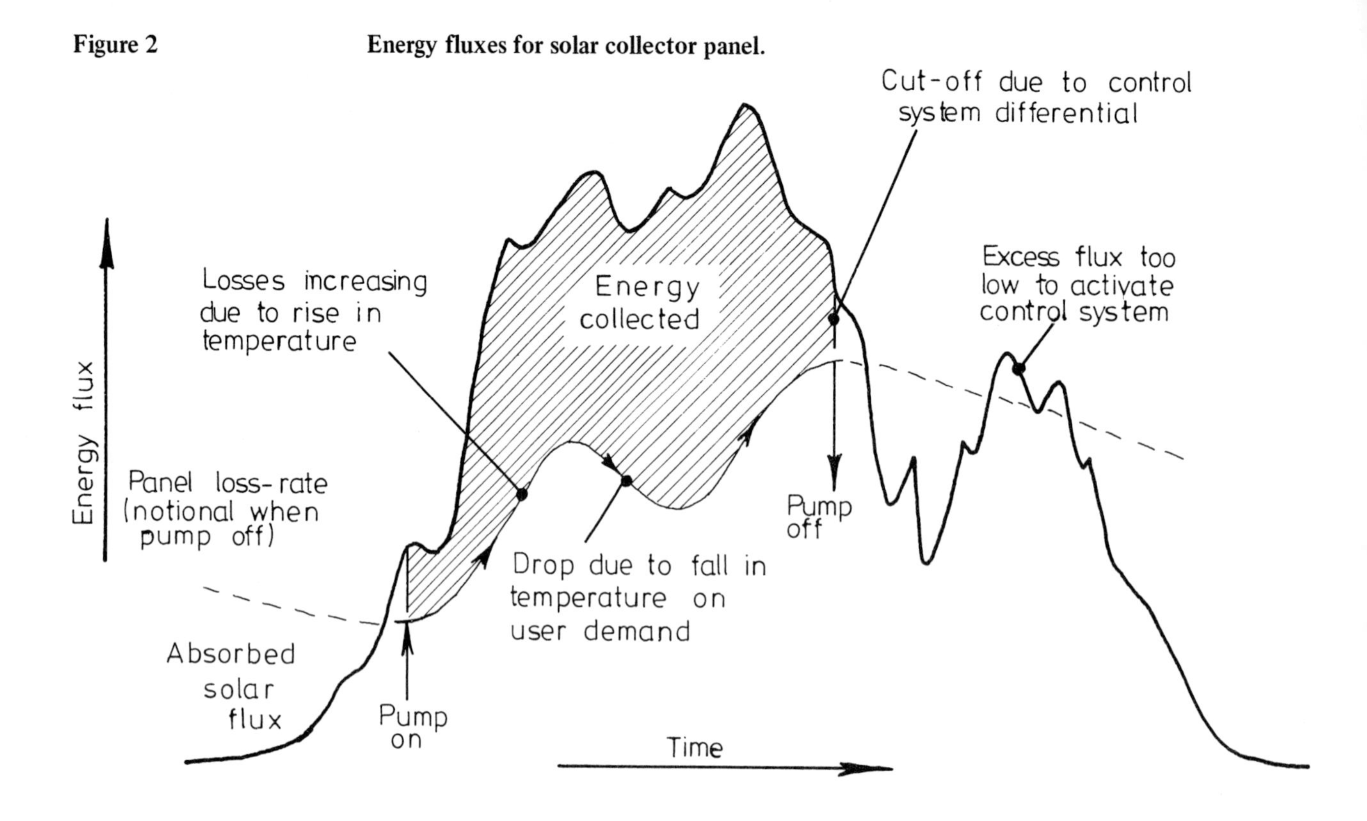

The temperature of water in the storage tank is determined by the previous history of irradiation and user demand. Though it will not be lower than the cold water inlet temperature, there will be days in winter when the solar energy flux does not at any time exceed the rate of energy loss that would occur from a panel, even at this temperature. It is evident that on such days as these, no energy gain is possible. The auxiliary energy required to raise the water to tap temperatures is then the same as for a non-solar system. On the other hand, there will be days in summer when the storage tank temperature has been raised above that required at the tap. No value attaches to that part of the energy collected which is responsible for the excess temperature, and a claimed annual collection efficiency should not include this. Moreover, these high temperatures are undesirable, as they may lead to boiling in the collector panels or scalding of an unwary user.

It can be seen that the behaviour of such a system is rather complex. Although it consists of only a few components, there is a complicated interaction between them in response to the variations in irradiation and user demand. These variations are specific to the geographical location and the life-style of the user, so that it could not be expected that experience in other countries would be a reliable guide to system behaviour in Britain. Moreover, if the behaviour were sensitive to the variations in user demand, even between similar households, this might make it difficult to predict the performance for a given installation in more than general terms.

These considerations suggest that the criteria of good cost-effectiveness in solar water heating systems will not themselves be particularly simple. The cost element includes the prime, installation, maintenance and running costs and factors such as the amortization period and ruling interest rates. The effectiveness might be quantified in terms of the value of the energy from conventional sources saved by the operation of the system. As we have seen, this is not simply the equivalent of the solar energy collected. Further, the reduction in cost to the user in operating the system depends upon the nature of the alternative system that would have been used in its place. At all but equatorial latitudes the solar system operates in the pre-heating mode described, and cannot be a complete replacement for conventional heating. Hence comparison has to be made with the marginal cost of operating what has been called the auxiliary heating arrangement. Only when this is an electrical system is the comparison straightforward.

Several assessments of the cost-effectiveness of solar water heating show the results to be critically dependent upon factors which are essentially unpredictable, notably future projections of ruling interest rates, rates of inflation and fuel costs (8, 9, 10). Whatever assumptions are made about these, however, it is clear that the need for good collection efficiency at low cost is paramount. Hitherto, it has not been established what features confer a good performance on the system as a whole. In the remainder of the present chapter, therefore, attention is directed at the technical factors which lead to minimum consumption of auxiliary energy. In particular, it will be the objective to see whether there are any configurations and operating regimes which render the system performance insensitive to matters such as the relative sizing of components and variations in the user's demand.

System modelling

Over a period of some years, a series of computer programmes have been developed, giving predictions of the performance of systems such as that illustrated in Fig 1. These are being used in system optimisation exercises, in which the leading parameters are systematically varied so as to show the dependence of the performance on each in turn. The first objectives have been to determine the effects of variation of collection area and storage tank volume for a given daily hot water demand. Variation in other physical parameters, such as collector loss coefficient and heat exchanger effectiveness will also be represented as time permits, but a programme which is a realistic model of the system is complicated, so that protracted and expensive computer running is required to explore the effects of even a few variables. This difficulty is probably the reason for the disconcerting lack of advice in the literature on primary factors such as required collector area and storage tank volume, though most other aspects of system design have been explored in minute detail (6).

To minimise the computation demanded, in the study now reported, the programme has been idealised in some respects. For example, though it permits variation in the representation of hot water demand throughout the day, this was restricted to a similar pattern on each day. The temporal variation in insolation is also idealised, so that only the main features are reproduced, without representation of short-period fluctuations.

It has already been shown that, in the case of indirect systems, optimisation of the collector/heat exchanger combination is possible to some degree, independently of the rest of the system (11). The procedure is simplified if it can be assumed that the water temperature in the storage tank is substantially uniform. Tests have shown that convective motions induced by heat-exchanger coils are sufficient to prevent the development of any significant degree of temperature stratification. It is then possible to represent the heat-exchanger characteristics in simple form. On the basis of tests with typical coils (7) a heat-exchanger effectiveness of 0.8 is used in the model. This permits an overall system effectiveness (11) of 0.85 to be achieved, with a fluid flow rate of about 0.025 kg/s per m^2 of collector area.

Collector characteristics appropriate to panels with non-selective surfaces and single glazing are assumed, so as to represent the majority of systems available in Britain at present. The model includes an algorithm for overall loss coefficient calculation which is perhaps unnecessarily complex. Tests with panels of the type assumed show that collector efficiency representation in simple characteristic form is probably adequate. A typical characteristic is shown in Fig 3, indicating that the scatter about a mean line, due to variations in such factors as wind strength and air radiation, may not be too serious.

The control system differential is taken to be 2°C and an anti-scald limit is assumed, such that circulation is stopped should the storage tank temperature exceed 65°C. Auxiliary heating is by electricity, with a thermostat setting giving delivery temperatures of 65°C when the storage tank temperature is below this. A regular daily demand pattern is assumed, and in the examples given here, 20% of the total demand is taken between 7 and 9 am, 20% around noon and the remainder after 5 pm.

A collector slope of 30°, with a south-facing aspect, is assumed. An approximate model

Figure 3 **Characteristic performance curve for typical single-glazed, non-selective collector.**

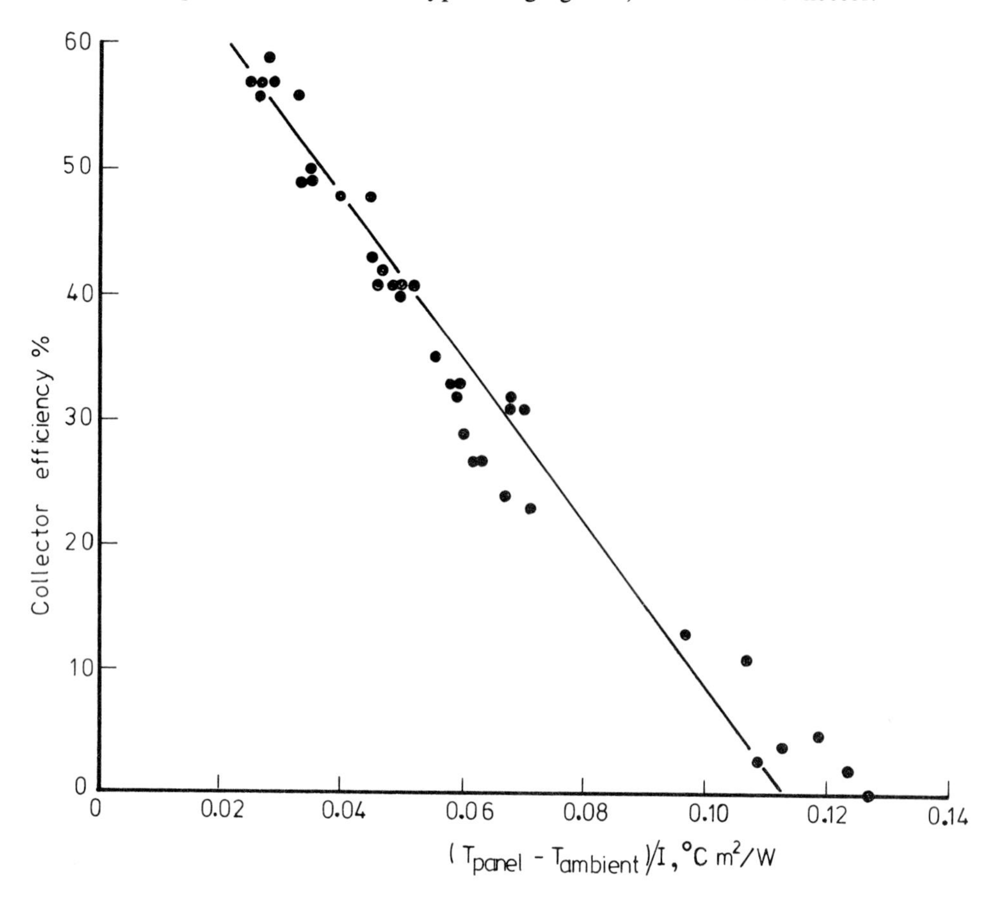

of the distribution of diffuse radiation over the upper hemisphere is in use, pending completion of a programme of measurements of the spatial and temporal variations in this. For the reflected radiation an effective albedo of 0.2 is assumed for surfaces within the collector field of view.

It is considered to be important that the day-to-day variation in insolation be properly represented. There are two reasons for this. Firstly, the behaviour of the system is highly non-linear with respect to the solar energy input, so that the average output over a period of time is not the same as the output under average irradiation. Secondly, since the energy collected on one day may in part be carried over to following days, the actual sequence of daily insolation variations can affect the performance. A method has been developed for quantifying the sequential characteristics of the UK insolation, based on the autocorrelation functions for the daily global insolation (12). For use in the model, sequences are employed which have the same internal autocorrelation throughout the year as has been found over an 8-year period for the global insolation at Bracknell. When the model was being set up, coupled data giving simultaneous variations in insolation and air temperature were not to hand, so that the ambient temperature representation does not have the same degree of refinement, being based on mean monthly values. No published data could be found on cold water supply temperatures, so that a measurement programme, now in its third year, was instituted for this and the resulting data used in the modelling.

The prediction programmes have been run for only a limited range of conditions at the time of writing. Some results of general interest concerning component sizing are given below. It is believed that overall conclusions drawn from them will be generally valid for systems operated in Britain. It cannot be said whether the right compromise has been struck between the degree of refinement with which a multitude of influences are represented and the time and cost of running which can be tolerated. This will vary with the amount of detail expected from the results. At present, only long-term effects are being studied. The validity of the model in predicting short-term behaviour is to be investigated by direct comparison with a real system which is being closely monitored.

System performance

The quantity chosen to represent overall system performance is the amount of energy required from the auxiliary heating source plus that supplied to the circulating pump. This is what the user has to pay for and is the quantity to be minimised, subject to acceptable costs. So that different systems may be fairly compared, this is reduced to a quantity A: the auxiliary energy required per unit mass of hot water delivered. This may be compared with the energy required to raise the water from inlet temperature to auxiliary vessel thermostat temperature, so that the effect of the solar pre-heating may be visualised. The difference will give the actual saving in heating costs to the user, though if it is to be reduced to a percentage saving, the losses from the auxiliary vessel, not included here, should of course be taken into account in both cases.

With the environment, demand pattern, collector/heat exchanger parameters and control logic all invariant, the system is characterised by the relative scaling of the collector and storage tank and by the magnitude of the daily hot water demand. These are expressed in terms of two parameters: the ratio R of the storage tank capacity to the daily hot water demand, and the quantity S, the daily hot water demand per unit of collection area. It is apparent, considering the quantities that are invariant, that S is a general measure of the ratio of the energy required for water heating to that available in the insolation upon the system.

Figure 4 **Monthly average of auxiliary energy required per unit mass of hot water delivered.**

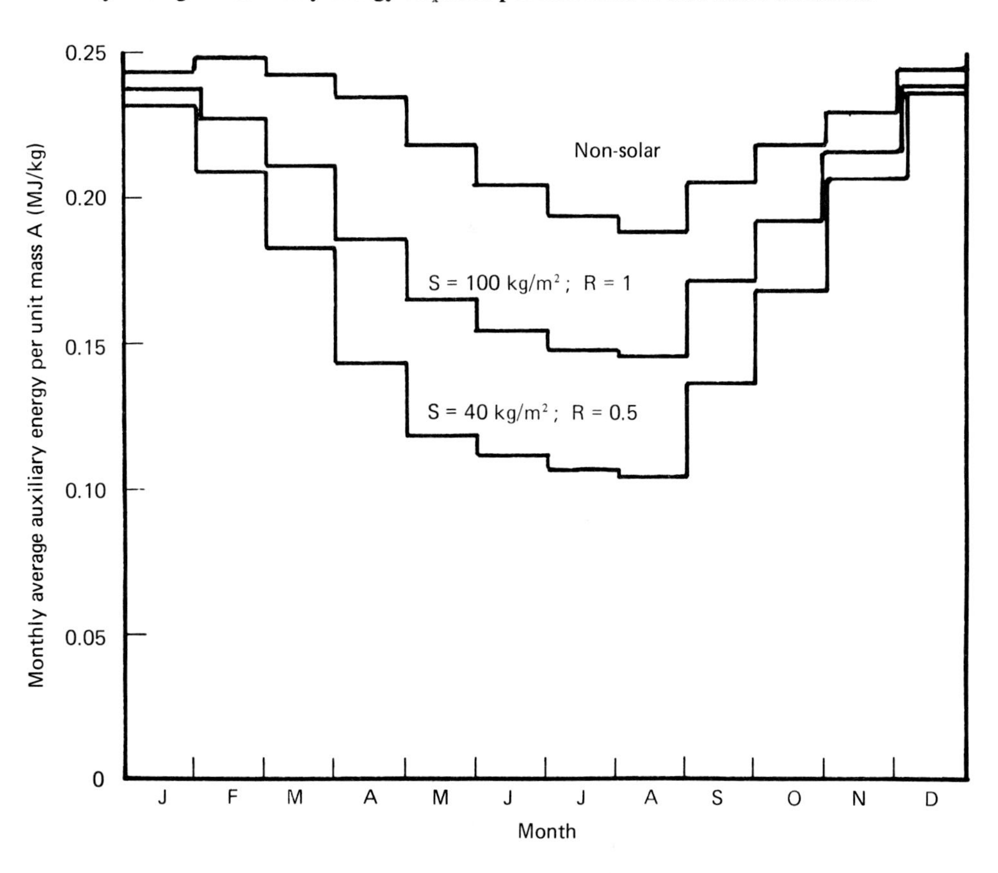

Some typical results, giving the monthly averages of the auxiliary energy required per unit mass of hot water, are shown in Fig 4. These, and a large number of others which have been obtained, give guidance on the sizing of components at different seasons of the year. For a daily hot-water demand which is invariant throughout the year, the annual average is significant, and in Fig 5 this is shown for a range of the parameters R and S.

Figure 5 **Annual average of auxiliary energy required, as a function of leading system parameters.**

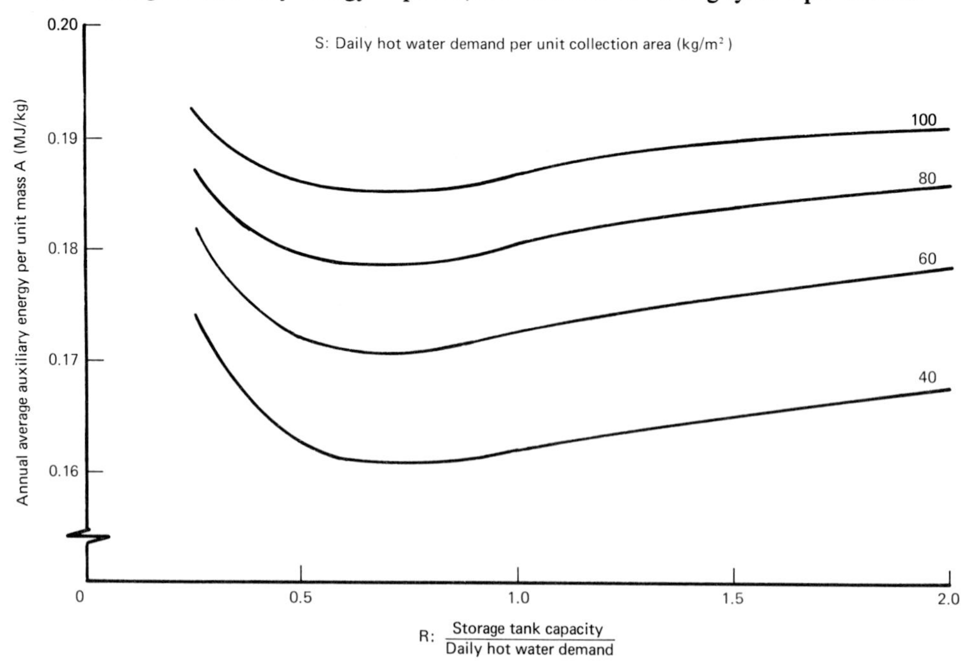

Results such as those of Fig 5 have important implications for system design. It is apparent that, with increasing storage volume, for a system which is otherwise invariant, the annual auxiliary energy requirement passes through a minimum. Fortunately, it is a shallow minimum, the system performance showing great tolerance as long as the storage volume is not too small. The indications are that if R is greater than about unity, the energy consumption per unit mass of hot water supplied is broadly unaffected by its actual value. This is consistent with the trend of the rule of thumb derived from Australian experience and reported by Duffie and Beckman (6), which suggests a value of 2.5 for areas with continuously good insolation, such as Darwin, falling to 1.5 for an area such as Melbourne, where the climate is less good.

The inward shift of the minima in Fig 5 with increase in S, suggests that the optimum values of storage capacity might occupy only a small range in absolute terms. At the minima, the values of the product RS, which represents storage capacity per unit collector area, lies between 30 and 60 kg/m^2 approximately. This would evidently provide a good basic design parameter for water heaters operated in Britain. It is interesting to note its resemblance to the 1 gallon/ft^2 used by Heywood over 20 years ago, on the basis of experimental work (13). This is about 49 kg/m^2.

The choice of the second parameter, which in practice would determine the collection area, is less clear-cut. As S decreases (collector area increasing for a given demand) the energy requirement falls, as expected, but not proportionally. It appears that the choice would be determined by economic factors, requiring an input of system cost per unit collection area. In practice, however, the situation is complicated by uncertainty about the user's demand for hot water. The IHVE Guide suggests an average of 70-140 kg/day per person (14), but it is known that there are great variations in practice. (Webster, for example, finds only about 50 kg/day per person for flat-dwelling families (15)). Fig 5 refers to the total requirement for the system and covers the range of 120-300 kg/day for 3 m^2 collection area to 320-800 kg/day for 8 m^2. Though no economic analysis will be attempted here it is pointed out that the results seem likely to favour smaller, rather than larger, collection areas. For example, suppose that a total demand of 240 kg/day is being served by a system having a storage capacity of 180 kg (ie R = 0.75). For collection areas of 3 m^2 and 6 m^2, the values of S are 80 and 40 kg/m^2 respectively. Then for the assumptions made here, we find that the annual average auxiliary energy requirement has values of 0.179 and 0.161 MJ/kg respectively. The energy required without solar assistance in this case is about 0.223 MJ/kg. Hence it is seen that a doubling of the collector area has increased the energy saving from 0.044 to 0.062 MJ/kg, that is, by about 40%. Plainly, the larger collection area would not be justified unless a doubling of area could be achieved with an increase in overall costs of less than this.

CONCLUSION

Limitations of space have made it possible to deal only with a few of the salient aspects of the behaviour. It is hoped that the dissemination of results of the kind given here will help to make more familiar the general features of solar heating systems and their characteristics, so that they may take their due place in the inventory of the building services engineer. It has been indicated that the performance is rather tolerant of variations in storage capacity, around a preferred value about equal to one day's demand. Alternatively, it appears that a storage volume of around 30 to 60 kg per m^2 of collection area would be an appropriate value for a design rule of thumb. A form of presentation of the results is suggested which enables a selection to be made of the collection area which maximises cost-effectiveness. Though no analysis of the economics has been undertaken here, it is indicated that these are likely to favour smaller, rather than larger collection areas for optimum systems.

REFERENCES

1. Brinkworth, B J, *Solar Energy for Man*, Compton Press (UK), 1972; Wiley (US), 1973.
2. Brinkworth, B J, Refrigeration and air-conditioning (in) *Solar Energy Engineering* Academic Press, 1977.
3. Szokolay, S V, *Solar Energy and Building*, Architectural Press, 1975.
4. Pillai, K K and Brinkworth, B J, *Storage of low grade thermal energy using phase change materials.* Applied Energy, **2**, 1975, 205-216.
5. Rogers, G F C, *Energy conservation – choice or necessity?* ChMechEngr, **22**, 1975, No 6, 65-69.
6. Duffie, J A and Beckman, W A, *Solar Energy Thermal Processes*, Wiley, 1974.
7. Brinkworth, B J, Some work of the Cardiff Solar Energy Unit. Proc Seminar *Solar Energy for Buildings*, North East London Polytechnic, Nov 26, 1976.
8. McVeigh, J C, Solar water heating in the United Kingdom – the position in 1976. *ibid.*
9. Courtney, R G, *An appraisal of solar water heating in the UK.* Building, **224**, 1975, 137-139;141.
10. *Solar Energy – a UK Assessment,* Int Solar En Soc (UK Section), 1976.
11. Brinkworth, B J, *Selection of design parameters for closed-circuit, forced-circulation solar heating systems.* Solar Energy, **17**, 1975, 331-333.
12. Brinkworth, B J, *Autocorrelation and stochastic modelling of insolation sequences.* Solar Energy, **19**, 1977, 343-347.
13. Heywood, H, *Solar energy for water and space heating.* J1. Inst Fuel, **27**, 1954, 334-347.
14. IHVE Guide, 1970. Inst Heating and Ventilating Engrs, 1971.
15. Webster, C J D, *An investigation of the use of water outlets in multi-storey flats.* J1. IHVE, **39**, 1972, 215-233.

DISCUSSION

Dr. D. Fitzgerald (University of Leeds)

Professor Brinkworth said that the main application of solar energy in the UK is to the preheating of domestic hot water and we hear about little else. We live in an industrial country and there are many industrial processes taking cold water from the mains or elsewhere and heating it. This is an enormous field to which all that Professor Brinkworth has said applies, except that there may be 100 to 200 m^2 of solar collection area per installation instead of just a few square metres. A pioneering installation for the preheating of industrial hot water was put in about two years ago in Nottingham. A collector about 200 square metres in area on the roof of the boiler house at the Raleigh Cycle factory is used to preheat boiler feed water and in a recent press release it was said that the average temperature rise was about 10°C. Similar installations could be considered wherever cold water is heated, in the textile industry, the food processing industry and in industry generally. I think industrial applications will be at least as important as domestic ones in the not-too-distant future.

Professor B. J. Brinkworth

The scale of industrial need for low temperature heating is very difficult to establish and can, I think, be misleading, because in many instances the low-temperature heating is obtained by condensing waste steam. Since the heat is available anyway, it is not a candidate for replacement by other sources. I mention this not to disagree with the point made, but to indicate that very often close examination of the system is necessary to establish whether or not the opportunity is there. Until recently, virtually no information was available on low temperature heating requirements in industry, but these are beginning to emerge from the Department of Industry's Energy Audit scheme which is accumulating a tremendous amount of information and it is only when this has been taken from a sufficiently large sample that it will be possible to assess just how real the opportunities are.

Professor J. K. Page (University of Sheffield)

The Australians have produced an interesting report on the possibilities of using solar energy for process heating. They identified the food processing industries as one of the most promising as a lot of water is used at a low temperature. If too hot, the food disintegrates. The water is normally produced through calorifiers and the temperature reduced.

J. Taylor (Atkins Research & Development)

It seems that the draw-off from domestic hot water systems has a marked effect on performance. It is obviously difficult to gather data on an average draw-off pattern, but has a survey been done, and is information being collected?

Professor B. J. Brinkworth

The performance is sensitive to draw-off and there is a lack of information. We have not been able to make any survey which is at all comprehensive. It is something that needs to be done and I hope will be tackled in the programmes which are now being established. We have made some measurements in my own Institution by installing solar water heating equipment in some student flats. We had to discover a totally unknown quantity, the amount of hot water used by a student. We set up a scheme for monitoring this a year or so ago and the data that have come out of it are very interesting. Information is not available and it needs to be if general predictions are to be made rather than predictions for a particular system in a situation which is being studied closely.

E. G. Brooks (BP Oil Limited)

Because of my personal experience, I cannot agree the reasoning on keeping down the size of the solar collector on domestic hot water supply installations. I put in three panels for solar collection, each $1 m^2$ in area, and found that I gained by adding a fourth panel. The pipework and all the control system was essential and completed for the three panels so the capital cost of the fourth was comparatively small, an additional 17%, and yet it provided 33% more hot water. When the sun is diffuse, less supplementary heat, provided by a gas boiler, is required. I am measuring the amount of fuel saved with the solar system by metering the gas used in supplying make-up heat for my fairly regular demand for hot water.

I am surprised that no mention has been made of the almost perfect load for solar heating, the open air swimming pool, or even the enclosed one, where the main demand is in summer, the requirement is for very low grade heat – the temperature of a large volume of water has only to be raised by a few degrees – and often pump and pipework is provided with the filtration plant. It can be installed, not for a small sum, but economically in comparision with the amount of heat saved. For solar heating to become more acceptable, it needs more economical installations to counter news of doubtful ones.

Professor B. J. Brinkworth

It is impossible to judge the example quoted without knowing more about the system and in particular the pattern of demand. I am not surprised by any result – it is interesting to have one at all on a system which has been carefully monitored. I suspect that if the result is examined closely, it will not be found difficult to reconcile it with what is said in my paper. Possibly one of the factors is that in my paper the index of performance was the energy consumption per unit of hot water delivered and that is not the sole cost to the customer. The customer has also to pay for losses from the domestic cylinder, which have not been included here, both in the reference system, the one without solar heating, and the one with it. Without having details of these it is very difficult to say whether it would be worth adding another square metre or not. In the example quoted in my paper, doubling the area would have reduced the required energy consumption by only 40% and would not have been cost effective.

The heating of swimming pools is a lovely application for solar heating – demand is in phase with the supply very largely and, because the collector is operating at low temperatures, a high collection efficiency can be obtained, very cheaply. It is not even necessary to glaze the collectors as a rule. But swimming pool heating is not a big part of the national energy consumption and perhaps is one that we might reasonably set aside in favour of some of the other, much bigger, loads.

R. Ward (Solarward Installations, Havant)

As an installer of domestic solar heating, I will offer some figures. A 33% increase in the panel area of an installation would add 15% to the cost as all that is necessary to increase the panel area from three to four panels is to put in an additional panel. The scaffoldings are up and the ancilliary pipework is there. I find it very unlikely that a three panel system on a four bedroom house would be more satisfactory than a four panel system, since the additional panel is so cheap in comparison with the total cost. Customers are not concerned with the difficulty of forecasting the performance of solar collectors and accept information that is offered. Unfortunately there are far too many cases of exaggerated claims for performance by salesmen who are sincere but misinformed. One has to offer some sort of a figure so tell customers that one square metre of collector could save £13 per year in electricity. It is a starting point. I should like to hear Professor Brinkworth's comments on this, perhaps in the context of what can be done about these claims in advertising, and on whether the Meteorological Office might provide authoritative information for those concerned with advertising standards.

Professor B. J. Brinkworth

On the question of panel area, I doubt that it is possible to reconcile an apparent conflict of views. If by increasing the area and thereby increasing the cost by 15%, we were sure that the collection would be increased by more than 15%, then it would clearly be advantageous. Misleading claims is a problem which concerns the International Solar Energy Society very much. We have all, I think, experienced people offering to save us more money than we are actually spending and we have all seen claims that would mean that the system was working at more than 100% efficiency. I think it would be very difficult for any kind of indictment to succeed under the acts relating to advertisements because it would always be possible to dispute the figures that were put against it unless they had an official stamp. You suggest that the official stamp might come from the Meteorological Office. It is a possibility, but it is planned that some sort of official stamp should come from the British Standards Institution. A committee of the BSI has been formed recently to establish codes of good practice for the selection of materials and for the design of systems for using solar energy and also to establish standard methods of test. To establish a British Standard test will take time and one needs to know what to do in the meantime. I cannot advise you on that. One needs to warn people against exaggerated claims and I think we can thank the media for doing this in recent months. On the other hand, we must recognise that the more of these cases there are, the lower will be the reputation of solar heating and that is very much to be deprecated. I can see no immediate solution.

Professor J. K. Page

As Chairman of the UK Section of the International Solar Energy Society I have felt I was on strong legal grounds in disputing a claim for collection when it exceeded the solar constant.

P. Warburton (Arup Associates)

From experience, load demands on hot water supply systems will change when it is known that solar energy is used to heat the water. Before a lengthy exercise, extending over years, is undertaken to collect data on present demands for hot water, it is worth remembering that they change when the installation is completed. We will not always do our washing on a Monday morning; hopefully, we will wait until the sun comes out.

Professor B. J. Brinkworth

That might well be the case but I think it would be unwise to assume that anything is likely to bring a big change of lifestyle. What we need to do surely is to develop systems which are tolerant of big variations in the load demand so it does not really matter what that load pattern is. Some systems are better than others in that respect. There may be some surprises in store for us. The characteristics of systems have yet to be fully explored. Relative insensitivity to storage volume, for example, is a very welcome feature from a practical point of view, but does not seem to have been reported on before. We have been looking recently at the sensitivity to other factors.

Professor J. K. Page

The water temperature data being obtained by Professor Brinkworth is just data for his locality. The Yorkshire Water Authority have pointed out to me that in Yorkshire water comes from three sources. That obtained from deep holes is roughly constant in temperature the year round, that from rivers is in the middle temperature range and that from upland moors is very cold in winter and is never very hot in summer.

6. The Application and Economics of Heat Pumps

J. Keable

Principles of Operation

In general, when a heat pump is referred to, it is an electrically driven vapour compression cycle machine that is meant. Fig 1 indicates the component parts of such a system, but it should be noted at the outset that the compressor may be driven by any means not necessarily by electrical power. The basic components consist of two heat exchangers, an expansion device, and a compressor. The system is sealed and filled with a refrigerant. Refrigerant vapour is compressed and in consequence is heated and pumped to the high temperature heat exchanger, where the gas condenses and heat is given out. The condensed liquid passes through the expansion device and cools before passing to the low temperature heat exchanger, where it takes up heat and evaporates before returning to the compressor, to renew the cycle.

This is the cycle in most refrigerators, and it is of course the cold end of the system which is of interest in this case. In its heating application however, although the temperature range involved is often quite different, the principles are identical. The heat source will usually be the surrounding ambient air, and the heat will be collected and emitted to the heating medium, usually either air or water. It is important to notice that the evaporator will be working at a temperature lower than outside ambient air; that the refrigerant will be selected such that its boiling point is also below that of normal air temperatures to be expected in the ambient air, when at low pressure; and that the evaporator is often referred to as the 'outside coil'. The condenser on the other hand must work at higher temperatures than those to be used by the heating medium, in order to account for heat exchange losses; the refrigerant will have been selected so that it condenses at a temperature higher than that to be used by the heating medium, when at high pressure; and that the condenser is often referred to as the 'inside coil'.

The vapour absorption cycle works in exactly the same way, excepting that the compressor is replaced by a heat absorption unit. The gaseous refrigerant is first absorbed into a suitable solvent such as water, if ammonia is used, and in this liquid state the pressure can be raised to the required level by a very small pump or by a simple percolating action. The temperature of the solution is then raised by the application of heat (such as a gas flame) at which point the refrigerant boils out of the solvent and enters the condenser as a gas. From then on the system is the same as for the vapour compression cycle, as after absorbing the low grade heat in the evaporator, the gas is redissolved in the absorption fluid.

Both these systems are capable of operating over a fairly wide temperature range, but the efficiency will tend to increase as the difference in temperature between the low grade "source" and the high grade "sink" deminishes. It is therefore important to ensure that everything possible is done in the design of the heat pump system to minimise the temperature difference normally operated.

APPLICATIONS

Heat pumps are often mentioned in the same breath as solar collectors, wind generators and similar ambient energy collecting devices. This should not mislead us however into the trap of regarding them as devices of the same order. Likewise, those who have become

Figure 1 **Component parts of an electrically driven vapour-compression cycle machine.**

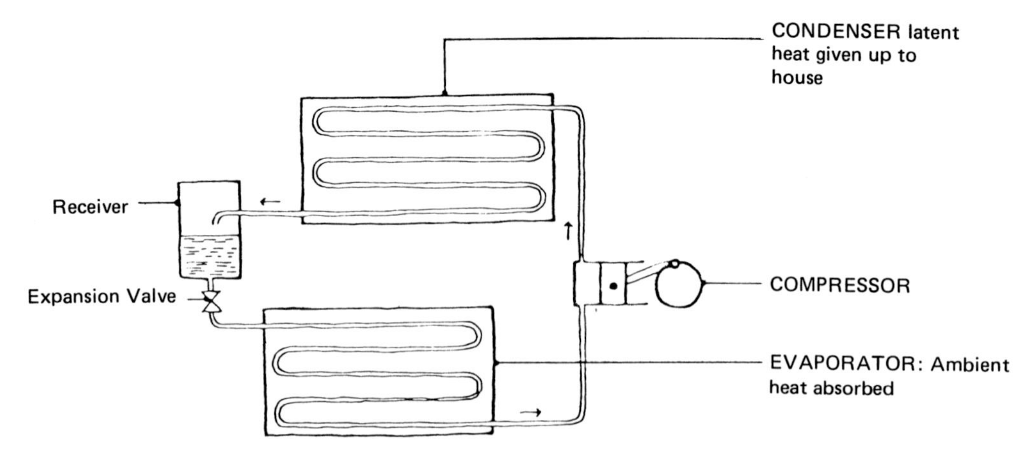

Vapour-compression cycle

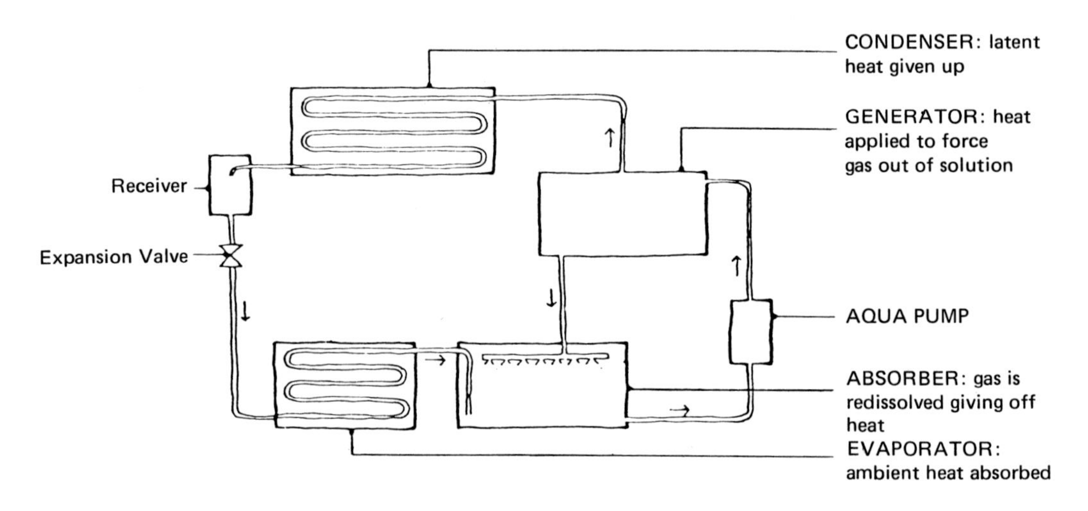

Vapour-absorption cycle

concerned by the inefficiencies of electricity generation and distribution tend to look askance at heat pumps, believing that they must necessarily be connected to an electrical supply. Although this is universally true of existing equipment, it is important to remember that in principle any energy source is capable of powering heat pump cycles, which as their name implies are simply devices for delivering heat to one point at the expense of cooling another.

The heat pump system was first proposed by Lord Kelvin in 1852 in an article *"On the economy of heating and cooling of buildings by means of currents of air"*. It has often been asked why a principle which is under ideal conditions capable of multiplying input energy by orders of magnitude, has not been more commonly applied up to now. Several reasons may be found, the principal being the inherent cheapness of combustion equipment compared with 'refrigeration' equipment (the principal factor so long as fuel remained cheap) and engineering problems particularly connected with compressor design at high compression ratios. The first of these reasons is very rapidly disappearing as fossil fuel prices rocket; the second reason was overcome to a very large extent by the work carried out in the 1960's in the context of reversible machines for the American market: machines capable of both cooling and heating a house as the season demanded.

If we turn now to consider the sources of heat for heat pump applications, we should remember at the outset that solar energy is the source of all ambient heat, whether it be the air, water or the ground that we are considering as our low grade source. In effect the atmosphere is acting as a gigantic energy storage mechanism, and particularly in the UK the temperature difference between the summer and the winter is to a very large extent softened by the presence of the adjacent oceans. We thus have in Britain an inherently suitable climate for heat pump application, since very low air temperatures are seldom encountered while soil and water are even less subject to violent changes.

Let us first consider at what points solar energy may be applied to a standard manufactured heat pump system, working on an air to air basis.

SOURCE (2) ➤ HEAT PUMP (3) ➤ SINK (1)

In the diagram the evaporator should be placed inside the source (in this case air) and the condenser would be supplying heat to the sink. Considering this diagram, we may see that solar energy can be applied directly to the point of use as in the Milton Keynes house for example: in this case the heat pump would be supplying the backup heating which in the Milton Keynes house is supplied by a gas fired boiler. The principle of operating flat plate solar collectors is widely understood, and the only point to notice here is that it will always be useful to minimise the amount of heating energy needed in a system by direct solar energy, if such a system is in any case going to be installed. However, space heating by definition cannot in general be supplied throughout the UK winter, unless it is linked with long term thermal storage.

Figure 2 **Schematic diagram of domestic heat pump installation.**

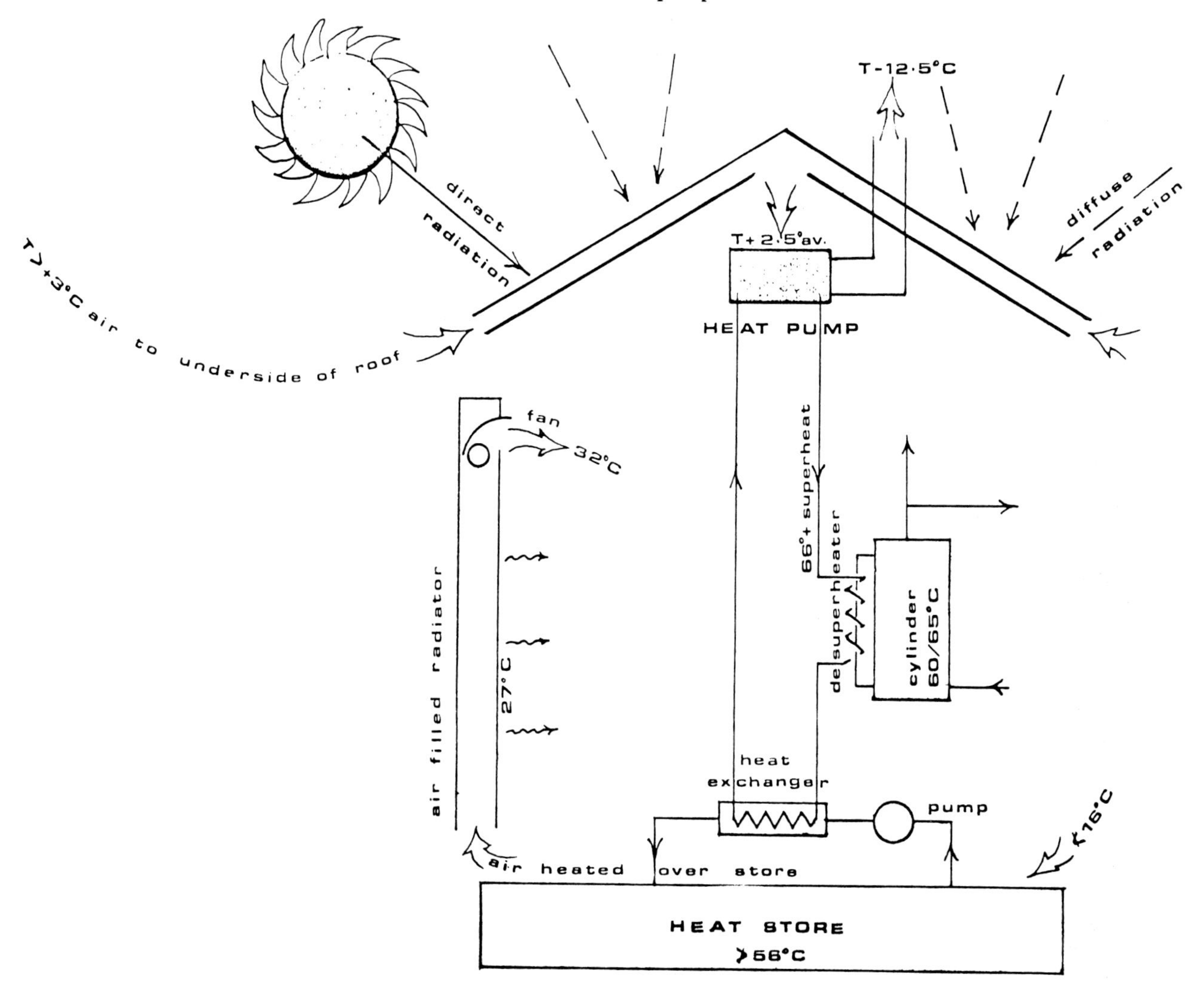

The second possible point of application is at source. This is extremely relevant since any reduction in the temperature lift required (ie a reduction of the temperature difference between source and sink) will be proportional to the heat pump efficiency. In the case of the air to air machine we are considering, it has been calculated that raising the ambient air temperature by as little as two to three degrees centigrade over the whole heating season may improve the efficiency of the system by as much as 9 or 10% (ie that much less input energy will be required).

A number of solar collector devices aimed at producing this effect have been proposed, but the possibility of using the fabric of the building to be heated, with only minor modifications, should not be overlooked. A normal slated roof for example will in itself act as a very efficient solar collector particularly where only very small differences in temperature are required to give useful improvements in efficiency. Tiled roofs will be slightly less efficient, but any building material will to a greater or lesser extent act as a solar collector, a fact which can be experienced in any attic during the summer months.

The Property Services Agency have examined the effect of loft heating in some detail, and have produced a ventilation system for use in their properties based on the fact that loft air is not only warmer but also less humid than ambient air, on average. This is of considerable importance in heat pump operation also, since humid air when passed through an evaporator will produce more ice given freezing conditions. In this connection it is worth noticing that our climate will particularly favour a low cost solar energy collector operating at a very low efficiency, since this will permit collection to take place from the whole of a roof, whether or not it is ideally oriented. The following calculation gives an example of a calculation for direct and indirect receipts in March, the indirect receipts being collected from a standard 30° pitched roof half facing south and the other north. The direct receipts will of course be from the south slope only, and as a result are only slightly more than half the total.

Direct receipts for March, on south slope only	=	$0.698 \times 0.62 \times 110 \times \sin(31^\circ + 30^\circ)$ kWh/m^2/month
	=	$0.4328 \times 110 \times \sin 60^\circ$
	=	41.63 kWh/m^2/month
Therefore mean daily receipts	=	36.91 kWh/day
Therefore during an effective 8 hour day	=	4.614 kW mean
Indirect receipts for March, on both slopes	=	$0.698 \times 0.22 \times 110 \times 0.5(1 + 0.8663)$
	=	15.76 kWh/m^2/month
Therefore mean daily receipts	=	27.950 kWh/day
Therefore during an effective 8 hour day	=	3.494 kW mean
Therefore total effective radiation	=	8.108 kW mean

If we now look at the application of solar energy at point three, the compressor or absorber as the case may be, these are some of the possibilities:

Electricity (= stored solar energy) – equipment available

Gas
Oil
Coal (= stored solar energy) heat may be applied to vapour absorption equipment at lower system efficiencies, but higher overall fossil fuel conversion efficiencies

Windpower (= solar powered energy transfer) could be used to power open compressor

Sun (= solar energy direct) would only be relevant in the cooling mode, applied directly to absorber.

At the UK Workshop on heat pumps, organised by the Energy Technology Support Unit at Lincoln College in mid-1976, possible application areas for heat pumps were reviewed. These included:

Domestic (both heating and domestic hot water, including the possibility of heat recovery from wasted baths, sinks and washing machines). Although it was agreed that in general air was the likely source of heat in most instances, both soil and water were recognised as having certain inherent advantages (particularly avoidance of temperature fluctuation), where these could be tapped without undue economic penalty.

Industrial: this was divided into three main headings firstly the use of rejected heat from refrigeration processes, for use both in water heating, process heat and pre-heating. Secondly, drying processes and boiling/evaporation processes. Thirdly, one off special applications where for any other reason heat can usefully be pumped from one process to another.

Social: in particular were noted schools, very suitable because the required high ventilation rates make the operation of a heat pump between out-going and input air feasible. Shops and shopping centres may be inherently similar to schools requiring high ventilation rates; possibility of using heat pumps to shunt between a thermal store (removing heat at times of high occupancy) and the reverse during the early morning warm up period. Swimming pools, both domestic and public, where very high operating efficiencies may frequently be possible for extracting heat from humid air.

Examples of Domestic Heat Pump Applications

The installation of standard air to air American heat pump machines has now reached the hundreds. Measured on a seasonal basis, as has been done by the Electricity Council Research Centre at Capenhurst, such heat pump installations tend to give an overall co-efficient of performance (COPh) in heating of approximately 2.2 to 2.4. This compares with the cost of off peak storage heating and as such is disappointing in terms of the theoretical possibilities; the equipment is relatively expensive as it has been designed for both cooling and heating (most people would agree that cooling is not required in the British climate) but reliability has been found to be good, that there has been no difficulty in obtaining draught free warmth with such systems (assuming normal duct grille sizing procedures are followed) but that defrosting does represent a problem in the UK climate (since the rate of ice formation is greatest when the air has a dew point and a dry bulb temperature both close to freezing: a relatively common condition here).

In mid-1975 the author together with Mr Christopher Dodson proposed a strategy of combining a standard machine such as those referred to above with the installation of a large thermal store and the collection of ambient energy by minor modifications to a standard roof structure. This proposal has been carried out in two instances; one in late 1975 as a conversion to the author's own house, and second an application to a new house carried out during 1976. In each case air has been directed onto the underside of the standard roof covering by forming an air passage on the underside of slates and tiles respectively so that solar energy both direct and indirect will be absorbed before the air reaches the standard heat pump machine. The heat collected from the heat pump is then passed to the thermal store directly, and the machine does not directly sense the indoor temperature of the house but only that of the store and of the ambient air. In this way it is possible to avoid running the machine when ambient air conditions are unfavourable (ie it is too cold) and this improves the efficiency of operation to a considerable extent, enabling COPs in excess of 3 to 1 to be obtained. The distribution of heat can then take place from the thermal store either by means of hot water to low temperature radiators or by means of air through a ducted system. In the case of the conversion the former principle was used; in the case of the new house ducted warm air was preferred.

Another interesting example of installation is the house built for the National Centre for Alternative Technology to a design by Peter Bond and Associates and built by Wates. In this case the requirements for heating were reduced to an absolute minimum by very extreme energy conservation methods, including 450 mm of dry-therm glassfibre insulation (producing a calculated U value of 0.0737 watts per m^2/degree centigrade) while double glazing was applied to double windows (ie four layers of glass in each). The front door was likewise double. These means it is claimed have reduced the energy consumption from some 2000 kW hours in the case of a conventional well insulated home to some 250 kW hours. The balance is supplied by a small heat pump of 1/5th horse power rating, and this is used for both heating and cooling: in other words for full internal temperature control. The outside coil is mounted in the roof in a duct, so that it does not in fact take advantage of the attic pre-heating effect but the arrangement does permit that air extracted from the kitchen can be diverted over the outside coil when necessary to prevent icing up in cold weather, and thus avoid the need for reverse cycling the machine to accomplish the same end. A second heat pump is used for heat recovery from wasted hot water, which in this house passes first into a storage tank under the ground before going to the drainage system. While the waste water is in this tank its heat is recovered and transferred to the hot water cylinder.

A further example is the house built by Ed Curtis in Rickmansworth in 1957. The house incorporates a 'solar wall' to trap directly as much solar energy as possible during the winter (heavy curtains are used at night to prevent the opposite effect, in addition to double glazing). After modification, the original air to air unit was modified to use domestic cold water as the low temperature heat source. This system has been operating since that time.

In other countries much more elaborate systems have been reported, particularly from the United States and Japan. In the latter case a solar house is reported from 1961 in which three heat pumps were incorporated, together with two basic storage tanks. The cool tank is heated by flat plate solar collectors on the roof; the heat pump then operates between this and the hot tank, and from here heat is distributed through low temperature radiation panels in the ceilings.

ECONOMICS

As a preliminary to discussing the economics of heat pumping, it may be useful to look at current fuel prices, reduced in each case to their equivalent in kW hours and hence to the number of units which can be purchased for £1. This is extended in Table 1 to show what £1 would yield using an electrically driven heat pump and an assumed seasonal coefficient of performance. It is interesting to notice that in the case of gas, using the inherently less efficient absorption cycle, the efficiency in terms of running costs or indeed in terms of national resources, proves to be very interestingly high. To date no hardware is obtainable following these principles, either for gas or fuel oil, though both are inherently practicable.

No figures are given for wind power, as the method of application would also have to be assumed. But it could well be comparable with off peak electricity. This table will be worth up-dating regularly as unit costs of the different fuels alter.

Energy Economics

At the Symposium on *Wind Sun and Waves* given at the Lorch Foundation in October 1976, Gerald Leach reported on the 'energy gap'. Many studies have shown, he said, that in the 1990's the gap between oil supply and demand will open, and it is normally assumed that this will be filled either by atomic energy or the relatively rapid development of other alternative energy sources. However, he went on to show that there are a number of possible situations in which the energy gap might open very much earlier, in some instances in the early 1980's! After examining the problem of providing the fixed capital formation required to meet the gap, he concluded that all conservation moves are worthwhile, if as a result electricity is saved and moreover, in view of the very short time span involved, he concluded that heavy investment was needed now in order to avert a resource crisis in the calculable future.

Seen from this view point (and granted that reducing energy demand at source by insulation, and the reduction of ventilation losses must in every case be the first objective) then the application of heat pumping techniques takes on a new significance. Provided heat pumps can be designed specifically for the British climate (that is to say optimised for our weather conditions rather than optimised for American cooling conditions, as at present) and provided heat pumps using different prime energies can be developed so that for example electrically driven heat pumps replace existing resistance heaters, while gas driven heat pumps replace gas fired boilers, then it would clearly be possible to push back the date at which the 'energy gap' opens even allowing for current estimates of energy growth

Table 1 **Current fuel prices.**

Energy	*Assumed unit cost*	*kWh/Unit*	*Cost per 100 kWh*	*kWh per £1*	*Assumed efficiency of fuel use*	*kWh yielded per £1*	*kWh yielded per £1 via heat pump*	*Seasonal COPh assumed*
Electricity (on peak)	2p/kWh	1:1	£2.00	50	100% resistance coil	50	150	3
Electricity (off peak)	1p/KWh	1:1	£1.00	100	100% resistance coil	100	200	2
Gas	14p/therm	29.3/therm	£0.48	210	60% in boiler	126	357	1.7
Fuel Oil	30p/ gallon	52p/ gallon	£0.58	170	50% in boiler	85	340	1.7
Coal	£20/ton	7350/ton	£0.27	370	40% in boiler	148	–	–
	Assumed cost of installation	Power output (seasonal average)						
Windpower	£3500 (5 kw m/c)	1 kw	£2.00*	50	100%	50		
Solar and interseasonal store	£8000	3.25 kw	£10.00*	10	n/a on a seasonal basis	n/a		

*No interest allowed; 20 year life maximum use.

rates. It has been estimated that taking all building types into account, some 50% of our total energy use is devoted to space heating. Assuming COPs varying between 1.7 and 3.3 it would not be difficult to save half of this amount which would represent one quarter of our total energy use. Even if it were only possible to arrive at a situation of 'no growth' of primary energy requirements, this alone would extend the life of our national resource, the North Sea oil field, by almost three times, according to Professor Alan Williams, Professor of Fuel and Combustion Science.

Turning to renewable energy sources, it is equally clear that if their effectiveness can be multiplied by a factor of two or three then a corresponding increase in their effectiveness as alternative sources ensues. For example in the case of wind power the maximum possible energy which can be extracted is 59% and 70% transferred as shaft energy: a total of 41.3% in all. If this can be multiplied by three rather more than the total energy available in the wind can be extracted from a given machine. The same situation pertains with water power, and wave power; the case of using solar energy directly has already been dealt with but one might add that the effectiveness of long term or interseasonal thermal stores supplied by flat plate solar collectors could be increased considerably by applying a heat pump to the energy extraction once they had fallen below their otherwise useful level. Depending on what operating temperatures are adopted, the size of such a store might well be halved: that is to say that by pumping the store down to a temperature just above freezing point as much heat could be extracted as had been stored above the usable level during the summer.

Money Economics

Heat pumps are inevitably more sophisticated than combustion devices. Up to now imported American machines have tended to cost something approaching £100/kW of heat delivered (capital cost). It would appear possible, assuming full optimisation of the machine to the UK climate (thus avoiding the complications of reverse cycling and much of the control gear) this figure could be halfed. UK Workshop on heat pumps recommended that machines capable of outputs in the range of 3 to 8 kW should be designed with system seasonal average co-efficients of performance of heating of three or more, and with a capital cost of not more than £250 greater at current prices than conventional heating systems. A second recommendation was that modular units should be developed with outputs of up to 2 kW, and at an installed cost of about £100, basically for individual room heating. These figures appear inherently reasonable.

Again looking at the problem in terms of overall resources, if we compare the capital cost of generating electricity, ie the need to renew or supplement existing generating resources, we may produce figures as follows:

Assuming 100 kW of heat is required as an end product

Example A: Generate 100 kW (delivered)
Use in resistance heater
At £600 per kW x 100 = £60 000 capital requirement

Example B: Generate 33 kW
Drive heat pump at COPh x 3 (yields 100 kW as required)
£600 x 33 = £19 800 + heat pumps at £50 per kW delivered £5000
Total capital requirement £24 800.

Of course this argument is even more compelling in terms of pure energy conservation (ie if a reduction in heat required is made) but is attractive even where this cannot be achieved.

IMPLICATIONS IN TERMS OF BUILDING DESIGN

These might be dealt with under several headings:

Energy Conservation:

All the usual arguments favouring energy reduction apply in the case of heat pumps as with any other sophisticated device. At say £50 per kW delivered, each kW not needed as a result of insulation, is that much saved in terms of both capital and running cost. The emphasis will undoubtedly continue to grow on reducing ventilation rates likewise and in the case of an air system it is possible and advantageous to have a fully controlled ventilation system with air change rates reduced in the case of domestic units to say one per hour.

Noise:

In the case of vapour compression cycle machines these should be sited as far from quiet areas as possible. In the case of air to air machines, large volumes of air must be passed over the outside coil and this requires relatively large fan sizes the main noise maker involved. This point will be particularly important if the machine is to run at night; much less so in cases where heating is only required during the day, or where thermal stores avoid the necessity of night time running.

If thermal storage is to be adopted then the siting of this will be important, assuming that water is the chosen material. Up to now it is certainly the most cost effective, but whatever material is used very adequate insulation is of course necessary. Moreover, aim to site the store below a zone where residual heat leakage will be useful.

Load Factors and Tariffs:

Where thermal storage is not adopted, then some form of supplementary heating must be supplied to deal with those very few coldest days during the heating season when the heat pump will prove undersized and will be operating inefficiently. Some alternative to electric resistance heating as a back up system must be devised if the electricity supply industry is to be attracted. The alternative is a situation where energy is saved most of the year, but the installed generating capacity has to be increased for peak demand only!

REFERENCES

1. *"Heat Pumps for use in Buildings"* BRE Current Paper CP 19/76.
2. Heap R D, *"Domestic Heat Pump Operation"* a paper presented at the Seminar on Heating and Ventilating and Air Conditioning at Imperial College London, June 1975.
3. Vale R and Vale B, *"The Autonomous House".* London: Thames and Hudson.
4. *"Wind, Sun and Waves"*, one day Symposium, The Lorch Foundation, October 1976.
5. Chapman P, *"Fuels Paradise"* London: Penguin.
6. *UK Workshop on Heat Pumps*, Lincoln College June/July 1976 Report published by Energy Technology Support Unit at Harwell.

DISCUSSION

W. R. Cox (Heating and Ventilating Contractors Association)

40 to 50 years ago, a general method of heating buildings was by low temperature warm water circulated through embedded pipes. A heat pump was used in a number of systems and the coefficient of performance approached 4 because the water temperature never exceeded 38°C (100°F) in those systems and water was generally circulated at around 32°C (90°F). The storage capacity of the building structure itself, with massive buildings, and the use of the low water temperature led to very good results. I see Mr Keable has returned to the idea of using the lowest possible temperature for heating in order to improve the coefficient of performance. I favour this because it produces a better method of warming anyway.

J. Keable

I am undecided on the relative merits of low temperature radiation (such as with buried pipes) and warm air. I tend not to like air heating systems because most of them are so bad, but I believe that a well designed air heating system may be able to operate on even lower temperatures than low temperature water in large radiant panels.

D. R. Macleod (Crerar & Partners)

Is the use of solar energy viable at latitudes greater than 55°? In particular, is it economical to use solar energy in Scotland?

J. Keable

The further north the location, the harder it is to make a heat pump viable, but the interesting thing is that in Sweden the experience seems to be that they are even more viable than in England. That sounds impossible, but the explanation is that when the heating season is longer and much more severe, although such a high coefficient of performance cannot be achieved, the overall effect – the nett gain – can actually improve. By extrapolation, I would say that Scotland is certainly a possible market.

R. W. Mattalon (Westinghouse Electric Europe, Brussels)

With a double roof covering to form an air passage on the underside of slates and tiles, as described, how is it possible to get heat without a "greenhouse effect" since we do not have a cover transparent to short wave radiation and an absorption in long wave infra-red? In my opinion, the solar energy absorbed by the slates would be lost by conduction and convection without helping the heat pump.

J. Keable

I cannot answer that question as satisfactorily as I would like at the moment. SRC have approved a grant to study the effects in detail. I have taken crude measurements though and I am satisfied that the rate of response is adequate. Slates are certainly better than tiles, because they are thinner, they remain dryer and on the whole they are blacker, but I would not like to say by how much they are better. Certainly the effect is there and is significant.

T. West (L. J. Couves & Partners)

In discussion on the economics of the use of ambient energy, many contributors have related the savings of the "free" heat obtained from solar panels, heat pumps and wind generated energy to the cost of electrical resistance heating. I feel this is rather an overstatement of the actual savings obtainable since electrical resistance heating is currently two to three times more expensive than heating with a gas boiler. This will obviously affect the current cost effectiveness of these systems by reducing the heat saving costs by a factor of one half to one third.

In the case of the heat pump installation in his own home, Mr Keable refers to the large water heat stores below the floor. Are these sealed to prevent condensation?

The requirements for storage of energy and the size of heat stores required for periods during winter when little ambient energy is available for collection have been discussed in detail but the figures are mainly based on weather statistics as the only available material. Mr Keable stated that his water heat stores had a storage capacity of about nine days, although he has only practical experience of the 1976/77 winter with his system. Did he find in practice that nine days heat storage was sufficient?

J. Keable

On electrical resistance heating, you are obviously right. From the figures given in the paper, you will see that the way in which electricity compares with gas and fuel oil is not entirely satisfactory and I think there is a future for heat pumps using other forms of energy than electricity.

When storing thermal energy in water, the water should be completely enclosed because losses through evaporation would otherwise exceed those due to conduction.

I would say that storage for a period of nine days is more than adequate, but only a monitoring programme would provide a proper answer. Our choice was based on ten-year weather data.

Moreover, the choice of energy source will be important, since there are fundamental differences in operating efficiencies between compression and absorption cycles. Assuming a reasonable pressure ratio in the machine, the compression cycle is capable of a COP of 3:1 or more. The absorption cycle, since it relies on a natural vapour pressure differential, has a limited range of "pressure ratio" across the system. The heating/cooling COP is thus less than one, but as there is normally the addition of heat to power the system, the COP in heating can rise to 1:1.7. In primary energy terms this may well be better than a compression cycle 3:1, since the latter normally implies using electricity, with the associated generation and transmission losses.

K. W. Reece (Ormrod and Partners)

The largest heat store is the earth itself. Heat can be obtained very simply from the earth by a heat pump for a house. Large quantities of heat can be extracted from the earth for district heating also. This country has many coalmines, which comprise a series of tunnels through which air is pumped to maintain comfortable temperatures and remove methane and other gasses. Should we not be thinking of pumping air in through old mines, many of them now sealed up, and converting to district heating by the use of very large heat pumps?

J. Keable

My own feeling about large scale schemes is, first of all, that they take very much longer to prepare. Secondly, the problem of control and payment are introduced. I think it has been demonstrated quite clearly that district heating is a disaster if people are not actually paying for what they take out of the system because they simply leave the windows open if there is more heat than they need. My view is that small units, like the traditional gas fired boiler, brought up to date by a completely different approach to the collection of energy will on the whole suit more people and prove to be more effective in conserving energy in the long run.

E. G. Brooks (BP Oil Limited)

The question of obtaining heat from the earth has received a lot of attention from the press and research teams have even investigated the granite rocks beneath Cornwall for their heat content, to see if cold water could be pumped down and collected a little warmer, but why spend money doing this when we are throwing millions of gallons of hot water away from every power generating station? Surely if we are wanting to recover low grade heat, we should take it from the known waste first.

J. Harvey (Ulster College, Northern Ireland)

My question concerns the location of the heat pump. I recollect from my student days that a good place to put it is in running water and I think of our island situation, surrounded by the sea with its very reluctant fall in temperature in the winter. I think also of river locations and references to remote island communities in Scotland. Mention of taking heat from waste domestic hot water makes me think of domestic wastes in total and the possibility of putting a coil of a heat pump in a domestic cess pit, which probably exists in many remoter areas, and extracting some of the bio-chemical energy from the decomposition of the sewage. Is this an idea worth considering?

J. Keable

The thought of a septic tank frozen solid is a bit daunting. Anyone who is lucky enough to be close to a water source certainly has a better heat source (in efficiency terms) than air. The emphasis on air collection arises because everyone has free access to it and not everyone has free access to water, but water provides a more stable heat supply.

J. Evans (Plymouth Polytechnic)

There is a problem with an air to air system of defrosting the evaporator coil. I take it that defrost arrangements are incorporated in your heat pump system. How often do you find it necessary to defrost under average conditions?

J. Keable

Both the installations referred to use American machines with automatic defrosting devices built into them. We have not experienced any problems and I cannot as yet tell you the frequency of defrost cycling. What I can say is that in a loft the conditions are significantly different from those in the outside air. There is a significantly lower level of moisture in the air and therefore frost builds up at a lower rate, but I am unable to quantify it.

Dr. R. W. Todd

The direct connection of wind machines and heat pumps was mentioned in the paper. Can any idea be given of what coefficient of performance might be expected if the compressor were driven mechanically rather than through an electric motor. My experience with small domestic electric motors suggests that their efficiencies are generally rather poor.

The negative aspect of using electrically driven heat pumps, which at present makes them less attractive than they might be, is low efficiency of electricity generation. One can argue that there is a case for heat pumps despite that, but I feel it is marginal. Looking to the future, larger proportions of our electricity may well be generated from ambient energy sources such as wind and wave power. Electricity will be obtained directly and efficiently from mechanical energy instead of by burning fossil fuels. The arguments against electrically driven heat pumps apply only when the electricity is produced from fossil fuels.

J. Keable

On a theoretical basis, an open cycle compressor could be used with wind power. Maintenance might be higher than with a semi-hermetic compressor, but I have heard nothing to suggest that the efficiency would be lower. If wind energy is transmitted through a hydraulic circuit, and the losses due to inefficiencies in the system, which are converted into heat, are available at the heat pump, I believe a coefficient of performance of three could be expected.

I would stress the wide range of application of the heat pump and believe its development is the key to many areas of ambient energy use.

J. G. Pym (British Petroleum Co. Ltd.)

Bearing in mind that most of the existing housing stock in the UK does not have warm air central heating and that it is supplied with single phase electricity, would you care to comment on the impact of heat pumps for existing houses?

J. Keable

I think the spread of the market might be fuel related. If there was a choice open to people to buy a gas fired or an electric heat pump system, I think there might be a tendency for people using gas to choose a gas system, and for people who already have night store or other electric resistance heaters to choose an electric heat pump.

The question of starting load is one of concern where only a single phase supply exists. It would be important to ensure with an estate of houses that a random situation exists and not one in which a time clock brings in units at the same time. For the mass market for existing houses it might well be that machines specifically designed for British (heating only) applications would include much smaller models which would be multiplexed. A full heating system would comprise several, just as several gas fires might be installed in houses. The problem of starting load is then reduced.

D. Taylor (Property Services Agency)

Has Mr Keable any information on coatings designed to disperse water forming on the evaporators of heat pumps, to reduce icing up?

J. Keable

I am aware of proposals for coatings but I think for the UK climate outside coils should be designed quite differently from the way in which they are designed for American machines. The American machines are basically cooling machines and are designed as such. For a heating machine, working in the winter with a severe icing problem, a more open coil may be appropriate.

Professor J. K. Page (University of Sheffield)

We are still lacking an enormous amount of data and an enormous amount of experience. Some of the work will obviously be done by the experts, but it is very important that the experts be reinforced by the professionally competent. We must build up a corpus of knowledge that makes the professions as a whole more able to contribute reliably in these fields. I feel it is important that the professionals do attempt to understand the technical principles better than at present, else ambient energy will become discredited at the start.

7. Low Energy Housing at the National Centre for Alternative Technology

R. W. Todd

INTRODUCTION

The Centre for Alternative Technology is a demonstration project, drawing together ideas and methods which contribute to a viable, sustainable alternative to currently accepted thinking in various technological areas – mainly energy, food and buildings. The permanent exhibition at the Centre attracts 40 000 visitors per annum. The bulk of this chapter describes the three buildings which form part of the exhibition but first some comment on the philosophy underlying the projects is necessary to justify some of the design decisions.

The alternative technology movement is seeking a technology and lifestyle which is sustainable in the long term, taking into account finite energy, raw material and food resources, the importance of fulfilling work for the individual and our impact on the ecosystem. It is clear that the era of cheap energy is coming to an end due to depletion of fossil fuel, while a large fraction of the world still has an energy consumption per capita less than ten per cent of Britain's. It appears inevitable that world population will at least double before any stabilising control can take effect and this increase will be mainly in the poor countries who are entitled to, and will ultimately demand, their fair share of the earth's wealth of resources. If we plan to live within our fair share, the long term future looks challenging but hopeful. If we continue to press for growth in our already excessive consumption the future looks grim. (1, 2, 3)

We can make an important contribution to the developing world and to our own future stability by conserving fossil fuels as far as possible for use as chemical feedstock. The most environmentally acceptable way of doing this is by massive energy conservation measures and the widespread adoption of ambient energy systems. Short term economic arguments are often used to reject such developments as solar heating; however, these arguments contain oversimplifications and must involve very uncertain guesses at the future to reach their conclusion. How, for instance, do you assess the 'value' of a solar heating system? Do you equate it to the cost of the energy it will save at current prices or try to guess the energy cost over the next twenty years? Or do you value it in terms of clean air, reduced nuclear risk, fossil fuels saved for more vital purposes in the future, or some security of energy supply in an uncertain future? It is clear that our motivation is rarely as simple as economists might suggest. Consider, for example, double glazing, which is judged by experts to be a poor investment in terms of energy saving, but is extremely popular nevertheless.

Decisions made on an 'economic' basis inevitably reflect the decision maker's view of the future and his own interests. Thus it is quite easy for a nuclear energy engineer to show that – with his value judgements – wind power is of little significance and likewise for the ambient energy enthusiast to put forward a convincing argument that nuclear energy is a sure pathway to environmental and financial disaster. I suggest that what really matters is to have a vision of the kind of future we want and to work towards that; economics is merely a tool which can help us to achieve it most efficiently.

In building design this is particularly important as we are designing for a lifetime of the

Figure 1 **A possible change in the UK energy demand and supply over the next 50 years.**

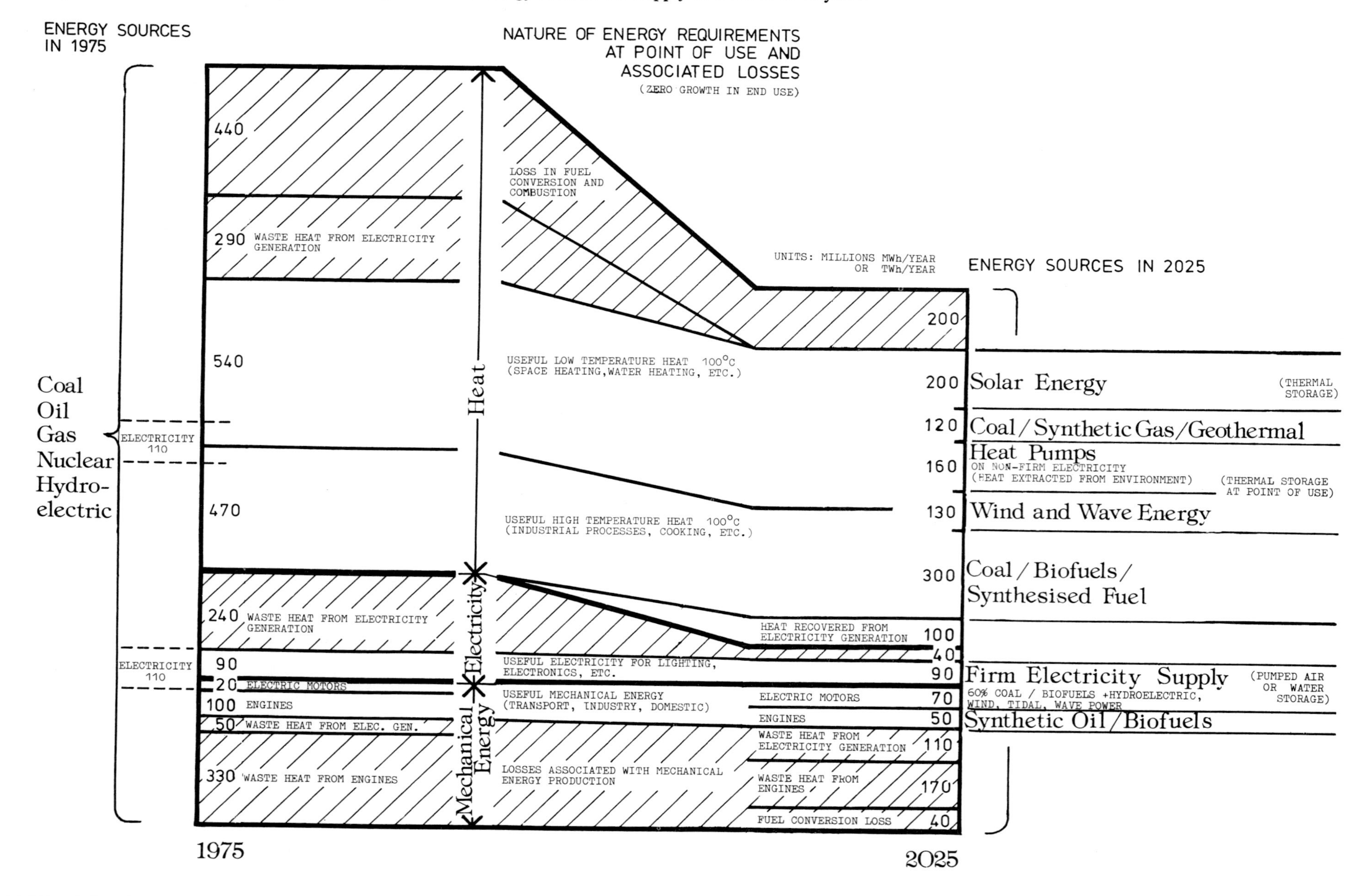

100 year order and economic guesses on this time scale are quite meaningless. Professor M W Thring states his challenging principle: *"What is essential for the long term survival of our civilisation is never economic and rarely politic."* However, short term cost-effectiveness is important insofar as it is a measure of efficient use of materials (but not built-in obsolescence), reflects good system design and encourages the early adoption of ambient energy devices. (3, 4, 5)

Our designs, therefore, take the use of insulation and ambient energy beyond the level generally considered economic and may also imply some reduction in comfort level below currently accepted design standards, but not necessarily below the comfort level in many homes at present.

WATES CONSERVATION HOUSE

In the summer of 1975 Wates were approached with the hope that they would take up the challenge of building a low energy house at the Centre as a demonstration of what could be achieved with existing building technology. Wates responded enthusiastically and commissioned Peter Bond Associates to design the house. During preliminary discussion the following decisions were made: to retain a building of fairly conventional appearance and size so as not to prejudice marketing possibilities, to limit the budget to £20 000, and to concentrate effort on reducing energy demand rather than supplying energy from ambient sources. The last point ensured that the house design would be suitable for any location, initially in most cases connected to mains services, but in isolated situations and perhaps more widely in the future, powered by ambient energy. The function of the house is partly to be a research tool and a means of gaining experience with large amounts of insulation and heat pumps but also to stimulate interest in low energy housing among the public, building societies, etc.

Energy is used in a house for temperature control, hot water supply, lighting, cooking and other electrical appliances. Energy required for maintaining a given comfort level in the building (in this case an internal air temperature of 18°C) is directly related to the overall U-value of the structure, the ventilation provided, and incidental solar and occupational heat gains. The design therefore employs a high degree of insulation in the walls, roof space and beneath the ground floor, small quadruple glazed windows, mechanical ventilation and fairly high thermal capacity inner walls to take maximum advantage of incidental gains without local overheating and to smooth out the effect of diurnal temperature variations. A heatpump supplied by Ebac Ltd is used for DHW supply and the design also attempts to minimise cooking energy and uses low power lighting. Rainwater from the roof is collected, filtered and used for all purposes except drinking, with automatic reversion to mains water when necessary.

Cavity wall construction is used; the outer skin is brick and the inner Thermalite block. The 450 mm wide cavity is filled with 'Dritherm' glass fibre which was inserted during construction with little difficulty. The outer wall is buttressed internally and tied only at buttresses to the inner wall. It is calculated that the Dew Point will occur on the outer surface of the insulation and since the insulation is layered in vertical planes, condensate will run downwards, together with any moisture penetration through the outer skin to the base of the wall where drainage holes are provided to the outside. The top surface of the ground slab is laid to fall towards the outer skin, discharging any condensate via the same drainage holes. A polythene vapour barrier is also provided above the ground slab insulation; this also serves the function of stopping glass fibre particles from being drawn into the building by the warm air heating system. The calculated U-value for the walls is 0.075 W/m^2°C.

A normal tiled pitched roof is used and to accommodate 450 mm of roof insulation without excessive building height, part of the ceilings of the upstairs rooms are angled. A polythene vapour barrier is provided beneath the roof insulation and the roof space above the insulation is naturally ventilated. Quadruple glazing brings the window losses more into line with the rest of the structure. A fixed outer frame carries one sealed double glazing unit and spaced at 200 mm, an inner openable frame carries a second similar unit.

Only one entrance is provided, thus avoiding through draughts, and this is fitted with double doors. Ventilation is provided by a 20 watt extractor fan in the kitchen, fresh air being induced into the house by the slightly reduced internal pressure. The air change rate is one quarter of the total house volume per hour and air is circulated round the house by the warm air heating system fan. Additional automatically controlled fan ventilation is provided for the shower room and toilet.

The calculated specific heat loss for the house is 66 W/°C. The peak demand is 1.2 kW and the space heating requirement with incidental gains taken into account, is only 8 kWh/day average in December and January and a total of 950 kWh over the heating season. This compares with 13 000 kWh over the heating season for an equivalent sized conventional house. The heat demands of the conservation house and a comparable conventional house over a typical year, taking into account incidental gains, are shown in Fig 2.

Figure 2 **Space heating energy demand.**

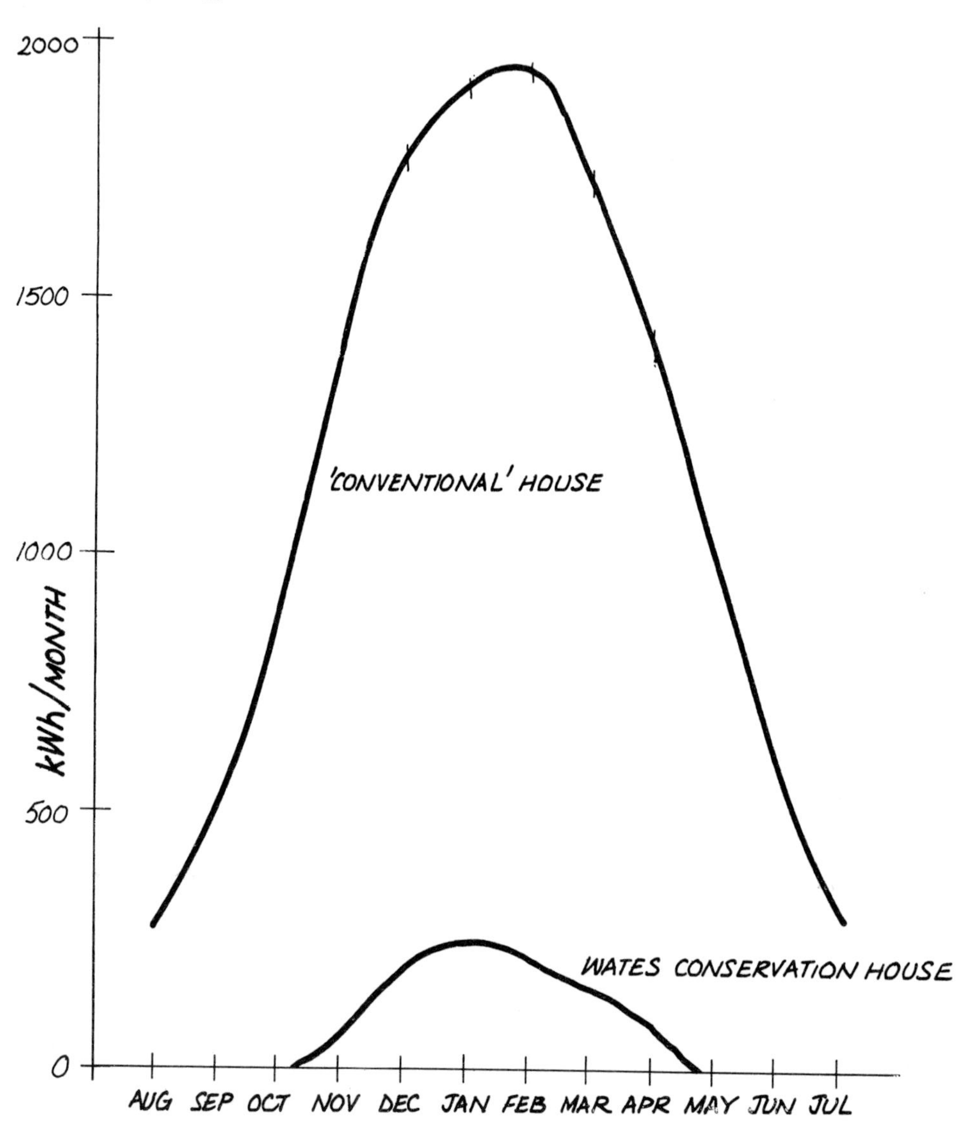

Heating system

An air to air heat pump was chosen for space heating and the low power requirement enabled a small and cheap system to be employed. The 1/5th HP compressor is the same type as used in a domestic deep freeze. The evaporator heat exchanger in the roof space is fed with outside air mixed with ventilation air from the kitchen extractor fan and the condenser is mounted with the compressor in a cupboard on the upper floor. Warm air is fed to all rooms via the under floor spaces and returned via ceiling level vents. The refriger-

Figure 3 **The Wates low energy conservation house.**

Winter Environment

Outside air is drawn through a vent in the roof by fan and passed over the heat exchanger. Heat is absorbed and the cool air is expelled colder. Collected heat is transferred by the heat pump to the core of the house.

Warmed air is circulated by normal thermal currents assisted by fans, through grills in the rooms, and under the ground floor. Some warm air is recirculated, mixed with fresh air to make up for air exhausted by an extractor unit in the kitchen.

This exhausted air, still warm, exits via the roof heat exchanger which recovers the heat in it. This also prevents the heat exchanger icing up in the very cold weather.

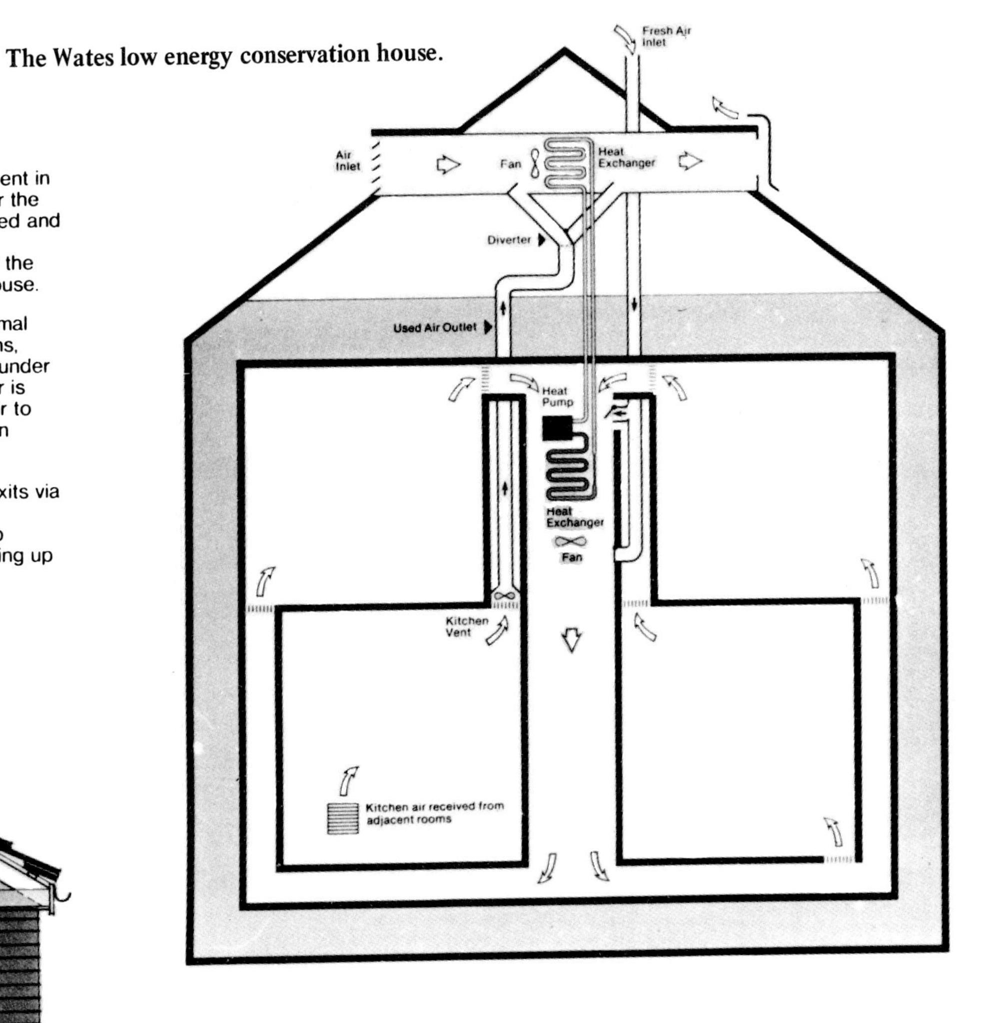

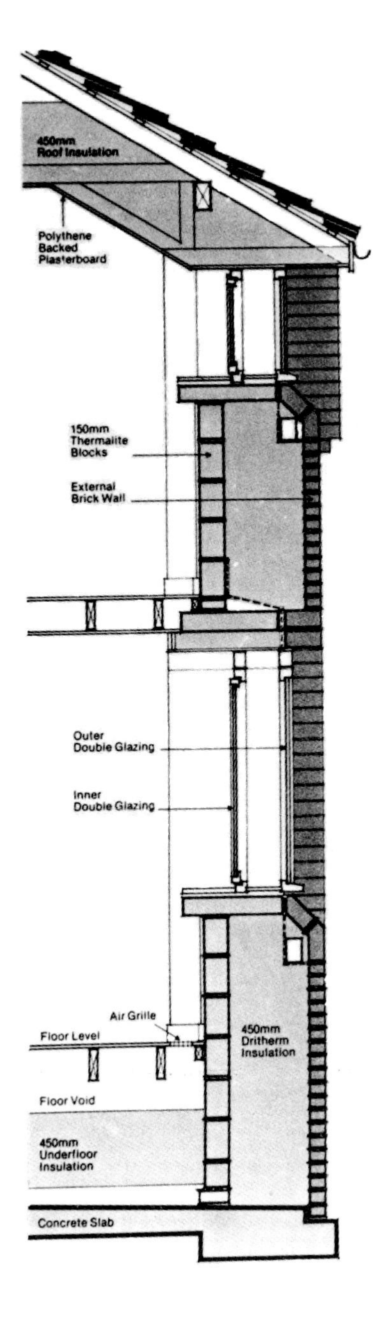

Insulation

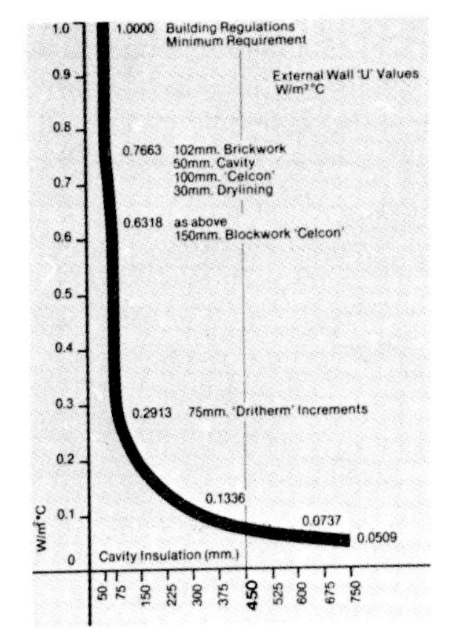

The low energy Conservation House was designed by Peter Bond Associates for Wates Built Homes Limited who donated it to the NCAT

To achieve the most effective heat barrier possible in the shell of the house: a cavity wall was designed incorporating 450 mm (18 ins) of Dritherm glass fibre insulation; the same thickness of insulation is used below the ground floor and in the roof; windows are double double glazed, i.e. two sealed double glazed panes with an 8 inch gap between them; the entrance consists of two separate doors; the cooking oven is enclosed in 150 mm (6 ins) of insulating material;

Substantial reduction of external influence on temperature inside the house made feasible the use of a small heat pump (1/5 h.p.) for internal temperature control.

ant circuit can be reversed by a changeover valve for automatic defrosting of the evaporator and for summer cooling should this prove necessary. Experience of the hot summer of 1976 suggests that in this location cooling will not be needed as the long thermal time constant of the house provides a very stable temperature with little diurnal variation. A COP of 2.5 is expected for this system.

Hot water

A similar small heat pump is used to heat the 40 gallon hot water tank. A large volume of relatively low temperature (50°C) water is used to maximise the COP while maintaining an adequate hot water supply (for most purposes cold water will not need to be mixed with it). The evaporator of this heat pump is immersed in a 60 gallon waste water tank below the ground floor. Waste water from the shower (no bath is fitted), all hand basins and one kitchen sink is fed to this tank and water from the bottom of the tank discharges to the drain as more water is added. By discharging water at a similar temperature to the cold water feed temperature, or colder if necessary, sufficient energy can be extracted to maintain the temperature of the hot water cylinder. It has been estimated that heat lost from the water to the house is approximately balanced by the electrical energy supplied to the heat pump. A COP of 2 has been measured for this system. Electrical consumption is estimated at 4.5 kWh/day, 1640 kWh/year, which is comparable with BRS estimates. (6)

Cooking

A well-insulated oven has been constructed based on a standard Belling oven surrounded by 150 mm of glass fibre. Tests have shown that energy use is reduced by a factor of about 2. An electric kettle and electric saucepans are provided as these are much more efficient than open ring cooking. So far tests are incomplete but we have successfully cooked a sponge cake with only 0.2 kWh! A modern version of the traditional hay box and an evaporative cooler are also fitted in the kitchen.

Energy supply

The house as described so far, with a total energy demand of about 1/5th of a similar sized conventional house, is suitable for normal mains electricity powering, but as the emphasis at the Centre is on ambient energy use and the site is not connected to the National Grid, a 3 kW Quirks aerogenerator was installed next to the house together with 20 kWh of lead-acid battery storage. It is anticipated that this will supply most of the energy required except for long windless periods combined with cold weather. A 1.5 kW standby generator is provided for such occasions. Electricity for lighting and cooking is supplied at 110V DC and a static inverter provides 240V 50Hz for the heat pumps and other essential AC uses.

It cannot be claimed that the wind/electric system is generally a cost effective option, mainly because of battery replacement costs. However, the initial cost of the system (about £3000) is similar to the cost of connecting isolated premises to the National Grid and in such situations a wind power scheme is much more attractive financially. However, a house designed specifically for wind power should not, of course, at present use batteries for the storage of heating energy because of their short life time and heavy replacement cost. Future battery developments may change the situation, but at present thermal storage is a much better option.

Present situation

The house is complete and has been open to the public for some months. The upper floor is partially occupied. Some preliminary tests have been made which so far appear to confirm design predictions but due to problems with the inverter, the house is not yet operating fully. A six channel temperature recording system has just been installed

WIND POWERED COTTAGE

In some situations wind power can offer advantages over solar power for the heating of buildings. Many existing properties are not well sited for solar heating systems and the space available for storage tanks is limited. A wind power system using thermal (eg water tank) storage requires considerably less storage volume to achieve almost complete independence from other fuels because:

(a) the availability of wind energy matches the heating demand quite well and the maximum storage time required is much less than for solar power which, for a practicable collector size, requires several months storage to achieve a similar performance (7), and

(b) the store can be operated up to much higher temperatures without loss of efficiency whereas high storage temperatures in solar systems imply increased collector losses and lowered efficiency.

Many isolated houses or groups of existing houses in areas with a fairly high mean wind speed (say > 5 m/s) could be heated by wind power. Such systems are, like solar systems, best combined with a high degree of insulation. The insulation is in itself cost effective in the short term, has a long lifetime, and should be the priority investment. It also reduces the size of ambient energy system required. An existing cottage at the Centre has been modified and fitted with a wind power system with the intention of providing a large percentage of the heating requirements, lighting and hot water for washing up and a shower. The system is shown diagramatically (Fig 4).

Figure 4 **Windpower system.**

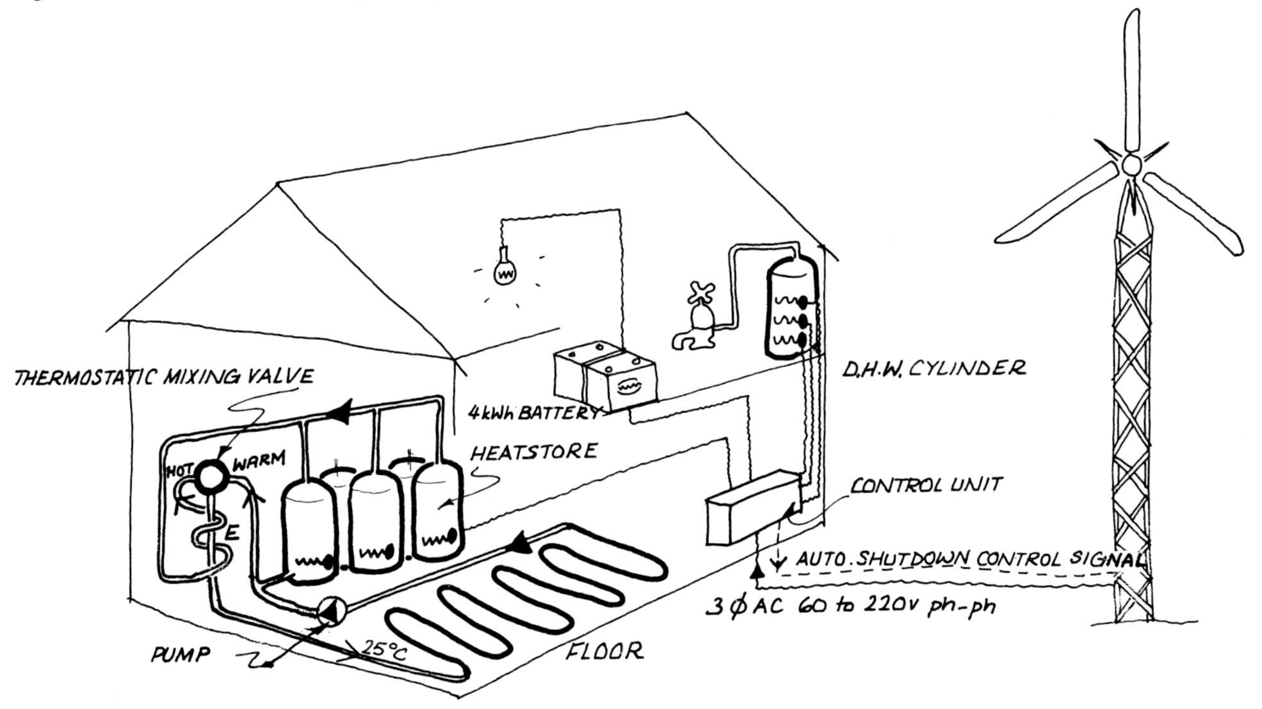

Building structure

The cottage is at one end of a row of three old slate-walled cottages and was in need of considerable renovation. It was decided to insulate internally using 75 mm glass fibre insulation covered by a polythene vapour barrier and plaster board supported on 3″ x 2″ studwork, giving a U-value of about 0.4 W/m^2°C. This was the maximum insulation level consistent with acceptable loss of room size. Roof insulation of 150 mm was also installed. Windows are only single glazed but are small in area. A new ground floor was necessary as part of the renovation and this gave the opportunity to install a water heated concrete floor. The 50 mm sand/cement screed is laid on 100 mm polystyrene block with a vapour barrier; 10 mm diam nylon pipes are laid at 150 mm centres in the screed to provide a large area low temperature heating surface.

Wind energy

The remainder of the design centred around the type of wind generator available. At the time the Swiss Elektro 5 kW machine was the largest well-tried aerogenerator readily available and the 3 phase AC permanent magnet alternator version of this machine was chosen. (A 10 kW version is now on the market). It was erected on an exposed site 250 m from the cottage, on an 8 m guyed tower. Meterorological Office data and some earlier on-site measurements suggested an annual mean wind speed of around 11 mph (5 m/s) which, assuming the usual form of velocity duration curve for Britain (8), and an average overall C_p* of 0.27 (the lower end of the range quoted by the manufacturers), indicated an average power output of about 0.8 kW could be expected. The manufacturer's published power/windspeed curves indicated a considerably higher C_p but the more conservative value is taken as this seems more usual for machines of this size. A recent article (9) suggests that these high claimed C_p values are not achieved in practice. Fig 5 shows the average power available for each month of the year, taking typical departures of monthly mean windspeed from annual mean windspeed and assuming that the same relationship between mean windspeed and average power output holds for monthly and annual means.

As the major part of the energy was required in the form of heat, thermal storage in water tanks was the obvious choice. The longest expected period of low windspeed (below the aerogenerator cutting in speed) during the heating season is the main influence in choosing storage capacity, but for reliable energy supply it is not sufficient to size the storage equal to the longest expected low wind period. A plot of the probability of the power available exceeding any fraction of the long term average power, P_a for various values of storage time is necessary to properly assess the optimum storage capacity. An

Figure 5 **Power input and demand for wind powered cottage.**

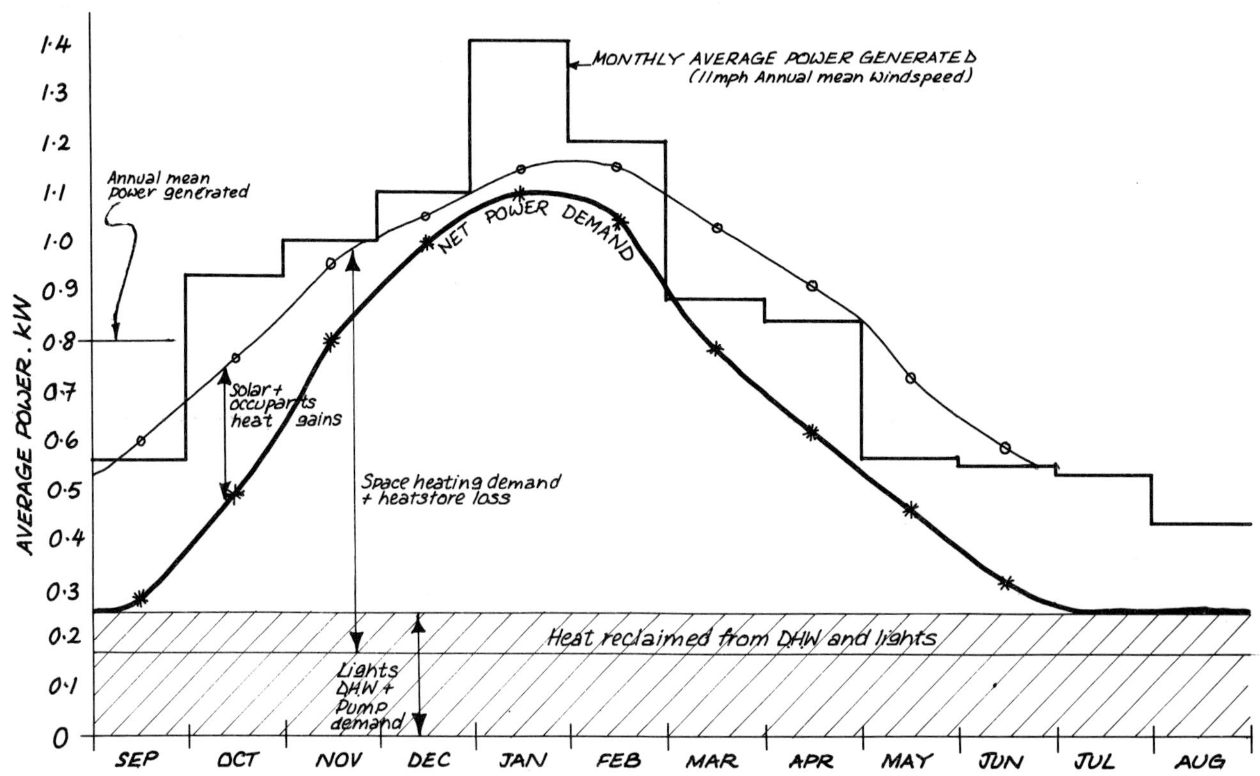

example of such a plot based on data given by Sorensen (10) is shown in Fig 6. Using such a diagram, the relation between storage time (the time for which the store can supply the load with no wind) and the ratio of load to P_a, for the required reliability of supply can be established.

In the example shown, eight days storage and load equal to P_a is assumed and the curve indicates a reliability of supply of approximately 90%. However this curve is for a particular machine and, with the lower cutting in speed of the Elektro machine compared to that considered by Sorensen the 8 day storage curve would be expected to rise to almost 1.0 as power available/P_a tends to zero. The problem is further complicated in a space heating application by the wide variation of load over the year which suggests that the plot should be made for each month of the year. It should also include the effects of variations in annual mean power, also shown in Fig 6.

$$^*C_p = \frac{\text{electrical power output}}{\text{power in the wind}}$$

Figure 6 **Wind power availability.**

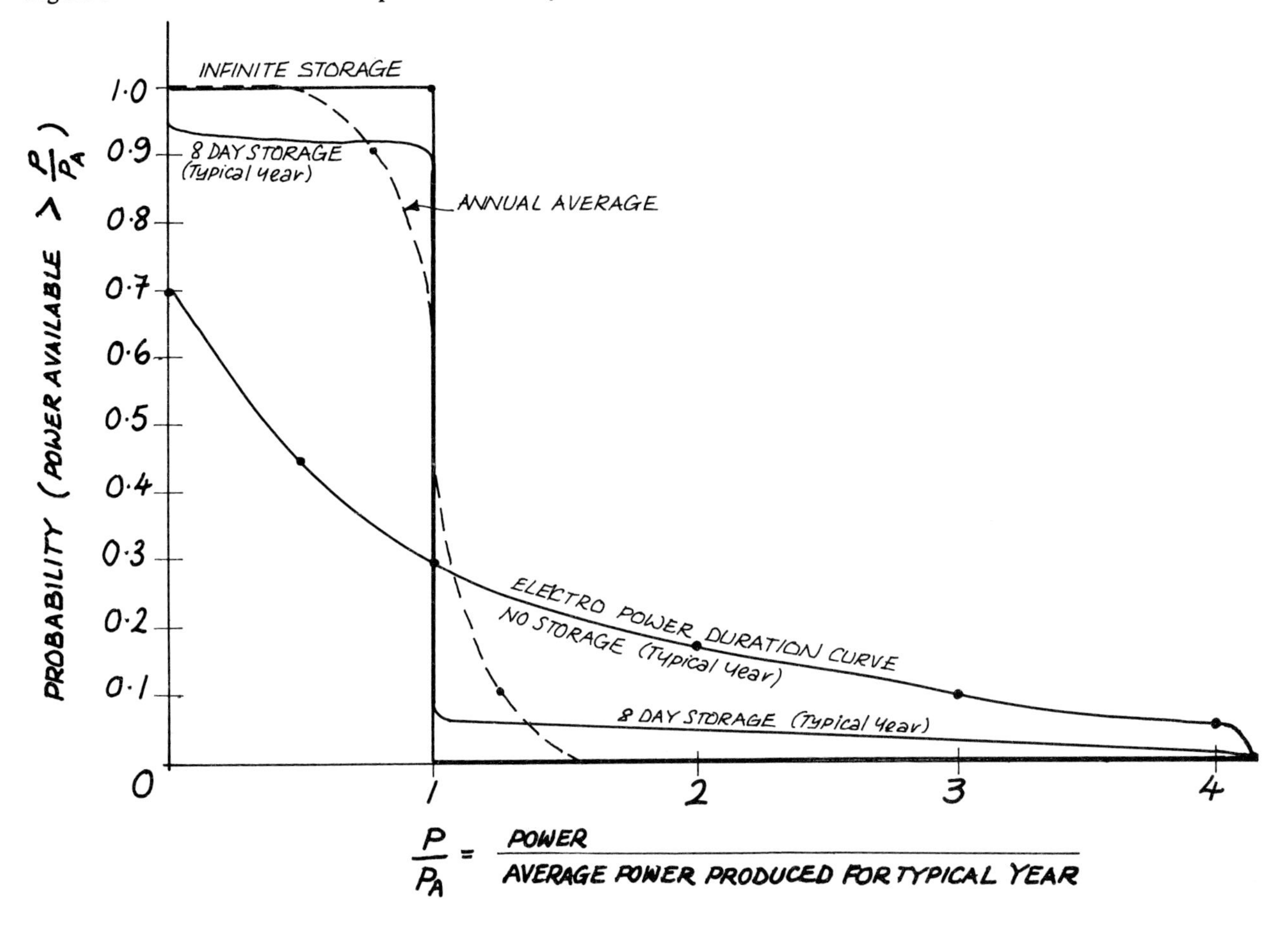

As inadequate data for this procedure was available for the site in question, a storage time (at midwinter load) of one week was chosen. The longest observed period with winds less than 7 mph – the Elektro cutting in speed – during the last two winters was about 4 days and it was decided fairly arbitrarily to choose a 7 day storage time to take account of the possibility of low average windspeeds either side of such a 4 day period. The storage time increases to about 10 days at each end of the heating season. This storage capacity is similar to that suggested by several other authors (8, 11, 12). The wind is being monitored on the site and further analysis will be performed when sufficient data is available.

Power demand

The power requirements of the cottage for lighting, and heating water, are minimised by using low power fluorescent tubes and localised lighting with low voltage incondesants, and by fitting a shower but no bath. The average power required for these functions is about 20 Watts for lighting and 220 W for hot water (17 gallons/day). A further 10 W average is required for a circulating pump. It is assumed that about 1/3 of the lighting and DHW energy can be reclaimed.

It was clear that the cottage could not be heated fully throughout with this size of aerogenerator so a compromise was reached in the design, heating the main room to about 18°C and the remainder of the house to about 12°C. The main room floor is heated directly and the heat store is positioned in the (ground floor) bathroom allowing heat to flow naturally from the ground floor to the upper floor. Because of the relatively low U-value of the outer walls and roof compared to the dividing floor, it is estimated that a temperature of at least 10°C should be maintained in the bedrooms. The heat loss coefficient for the main room, including ventilation and loss to other rooms is calculated to be 56 W/°C and the average power demand for continuous heating is shown in Fig 5, together with the estimated average power produced per month and the lighting and water heating demand. Although this suggests that the design is satisfactory, it must be remembered that the power inputs and demands quoted are long term averages for a particular month and there may be significant variations between the same month in different years.

However, there are various energy economies which can be made to meet such variations, mainly on the space heating demand, such as reducing temperatures all round, reducing or using heat reclaim ventilation or more simply heating for only part of the day. Although the floor screed will have a cooling time constant of about 8 hours, a significant saving could be achieved by only heating for part of the day. Such economies will probably be necessary as recent wind monitoring on the site suggests that the winter windspeed is significantly lower than the original estimates and as energy produced falls rapidly with decreasing windspeed, a fall to say 9 mph would amost halve the energy produced.

Heat store and controls

The heat store consists of five 60 gallon copper hot water cylinders, each fitted with an immersion heater. These are thoroughly insulated with about 150 mm of glass fibre. They operate under 4 m of water pressure head and the working temperature range is 25°C to 105°C giving a storage capacity of approximately 130 kWh. The tanks are coupled in parallel and feed a thermostatic mixing valve to ensure a water temperature of around 25°C is fed to the floor with any tank temperature above 25°C. A heat exchanger (E in Fig 4) is necessary to limit the maximum hot water input temperature to the mixing valve.

The immersion elements, together with three more immersion elements in the domestic hot water cylinder are switched in various series/parallel arrangements to match the load to the power available as the wind fluctuates and to give priority to domestic water heating when necessary. The switching is achieved by electromagnetic relays in the Elektro control unit which has been substantially modified for this application. A 400 W battery charging circuit feeds 4 kWh of lead-acid battery storage for the lighting, pump and control operations. Room temperature is controlled by thermostatic switching of the circulating pump.

Present situation

At the present time the system is complete and has been operating but is not yet satisfactory due to several problems with the aerogenerator and its control system. The problems centre around ensuring optimum matching of load and generator. With the recommended loading, the automatic high wind shut down mechanism, triggered by too high an output voltage operated far too frequently necessitating manual resetting each time. Modifications are in hand to avoid this problem and to provide automatic resetting. When these are complete comprehensive tests will be made on the whole installation.

SOLAR HEATED EXHIBITION AND OFFICE BUILDING

The building is a conversion based on an old slate-walled workshop and now houses an exhibition hall and bookshop, with an office above, totalling 200 m^2 floor area. To achieve a low specific heat loss, an internal perimeter wall has been added and the building has been insulated with between 150 mm and 200 m of polystyrene all round. With double glazing, window shutters and a low ventilation rate, the specific heat loss is calculated to be approximately 250 W/°C. The heating system design attempts to provide all year round solar heating with no auxilliary input. It consists of a double glazed 100 m^2 trickle type collector based on Thomason's design but incorporating some improvements, a 100 m^3 water heat store and the ground floor sand/cement screed which provides a large low temperature heating surface. Air circulates naturally between the two floors, heating the office. A low internal air temperature of 15.5°C has been chosen for the winter months but the large radiant floor area at approximately 23°C and wall temperatures almost equal to air temperature should provide an acceptable comfort level.

Calculations based on the incomplete radiation data available suggest that on a monthly basis, energy collected should meet the heating load up to the end of October and from early February. The main area of uncertainty is determining the value of collector efficiency to assume in midwinter as the insolation is often near to the critical level below which useful temperatures cannot be achieved and the use of average values of insolation, even hourly averages, can lead to considerable errors, as the collector response time is quite short. However, during December and January, most of the heating load is to be supplied from storage so the effects of inaccuracies in calculating collected energy for these months is small.

Collector

The collector uses factory coated (non-selective) corrugated aluminium sheets with aluminium glazing bars and ridge pipe. Glazing is 2 x 4 mm glass in 1930 mm x 580 mm sheets. All the remaining pipework in the system is aluminium or plastic to minimise corrosion problems. The collector back insulation consists of 150 mm of 'Nilflam' polyisocyanurate block to withstand the high temperatures sometimes reached in no flow conditions. An air space between the absorber and the back insulation which is normally blocked top and bottom can be used to dissipate heat by natural air convection in extreme situations such as pump failure coupled with prolonged high intensity solar input. The water flow rate through the collector is 0.4 litres/sec, a compromise between achieving low thermal gradients and minimising pump power consumption – which must of course be powered from ambient energy on this site.

Figure 7 **Solar heating system diagram.**

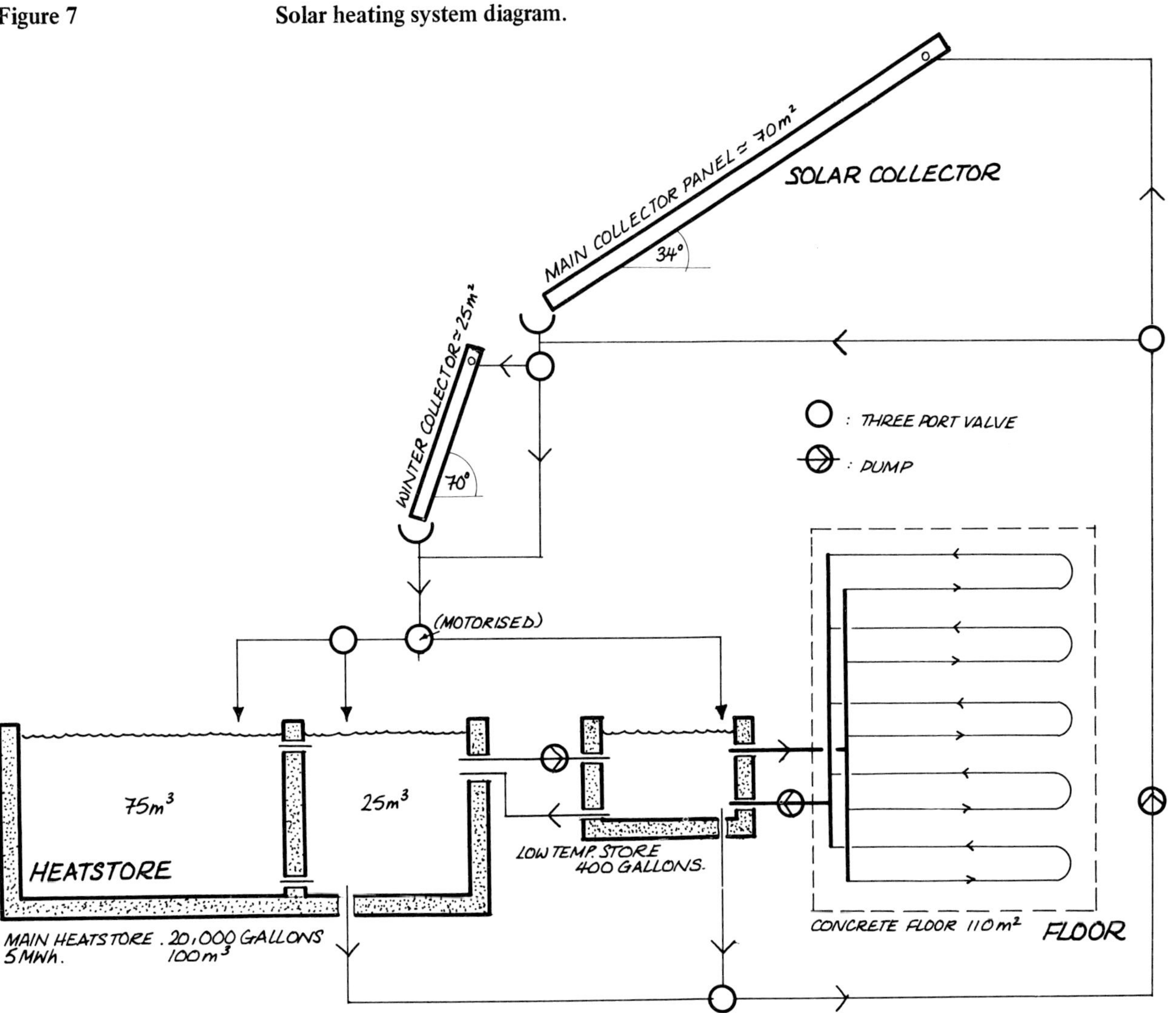

Heating system and store

In order to make the most of low intensity radiation in winter, the water temperature required for heating has been minimised by using the 110 m² floor screed as the heat exchange surface; Nylon pipes 10 mm diameter at 200 mm centres are laid in the screed which itself rests on a vapour barrier and 150 mm of polystyrene. Floor temperatures in the range 18 to 25°C are anticipated depending on heating load. At the design stage it was considered that stratification in the main store could not be relied upon over the long storage times involved, so in order to benefit from the low temperature heating system a small low temperature store is provided separate from the main store. The control system maintains this tank in the 25-35°C range to enable low intensity radiation to be collected

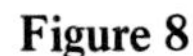

Figure 8 **Annual cycle of heatstore.**

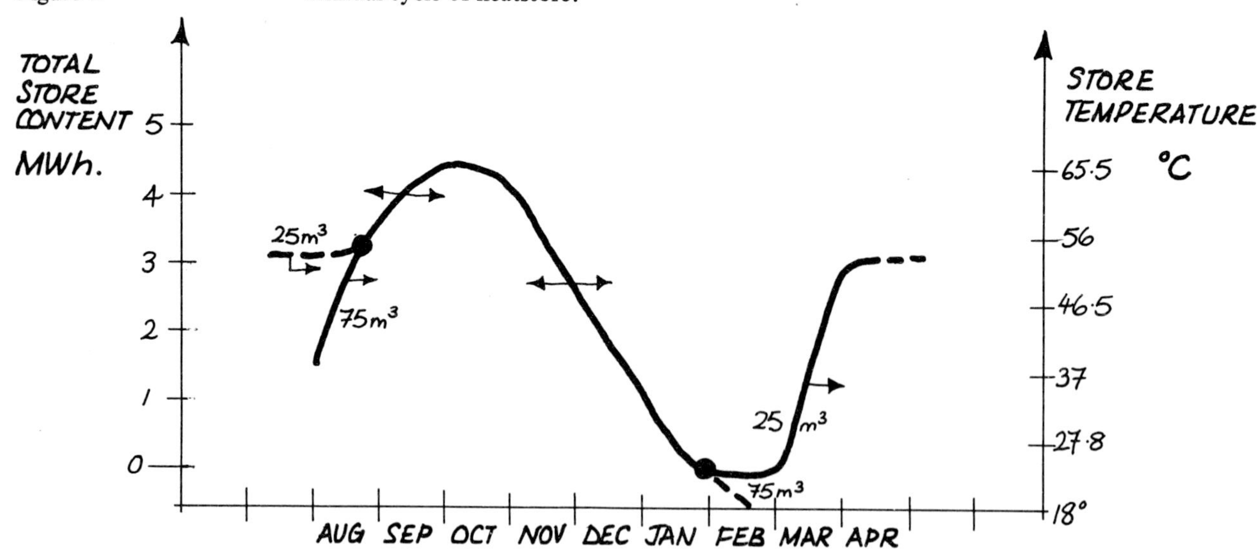

and used at low temperatures. This allows efficient collection even when the main store is at a high temperature in the period between summer and midwinter. The system interconnections are shown in Fig 7.

The main store is segregated into 25 m³ and 75 m³ so that the thermal losses of the whole tank need not be carried after January when its very large thermal capacity is not needed. A single valve between the two parts is closed when the main store temperature has dropped to its minimum useful value (probably early February). The 25 m³ section with its much reduced heat loss is then used for the remainder of the heating season and the whole tank is re-heated in summer (Fig 8). The control system employs semiconductor temperature sensors and an integrated circuit logic system to determine when to run each of the pumps and to control the motorised valve.

The main store was constructed by excavating a hole in the ground of "V" section so that the walls were quite stable and lining this with 300 mm thick polystyrene blocks sealed together by pressure injected polyurethane foam followed by a loose fitting butyl rubber liner. It has been found necessary to line the corners of the tank beneath the butyl liner with matting to prevent the liner ballooning into small cracks in the insulation which inevitably appear due to settling. The top insulation is provided by floating 300 mm polystyrene blocks with a vapour barrier. The present tank is outside the building for ease of access, but could be below the building in future designs with some thermal benefit, if a good alternative to the low temperature floor heating system could be found, and the additional heat gain to the building in summer was acceptable.

Present situation

The building is in use, the collector is complete and has proved satisfactory in tests, but work is still progressing on the heat store and control equipment. Thermal tests on the building indicate that its heat loss is close to that calculated.

CONCLUSION

It is anticipated that the three buildings described should yield useful data over the next few years and also serve to stimulate interest in low energy and ambient energy buildings. Experience in operating ambient energy systems at the Centre will help to answer more fully the many requests which we receive for advice and will point to areas needing further research and development.

Looking to the future, it appears that significant initial cost and raw material savings in solar and wind powered buildings could be achieved by building groups of houses which share a common solar heating system or a large windmill. In the coming year it is planned to install a 150 kW Wesco wind turbine at the Centre to provide the remaining buildings with space and water heating employing direct mechanical to thermal conversion, a common storage tank and hot water energy distribution along the lines of a small district heating scheme. Such a scheme combined with low energy houses holds considerable promise for out-of-town developments, one machine powering about thirty houses.

REFERENCES

1. The Ecologist, *Blueprint for Survival,* London: Penguin 1972.
2. Schumacher, E F, *Small is Beautiful*, London: Abacus 1974.
3. Thring, M W, *A World Energy Policy*, Inst of Fuel 'Potential for Power' symposium at Southampton University, January 1977.
4. Chapman, P, *Fuel's Paradise*, London: Penguin 1975.
5. Working Party Report: *Energy Conservation* – a study of energy consumption in buildings and possible means of saving energy in housing, CP56/76 BRE 1975.
6. Syemour-Walker, K, *Low Energy Experimental Houses*, BRS Energy Research and Buildings, 1975.
7. Szokolay, S V, *Solar Energy and Building*, London: Architectural Press, 1975.
8. Golding, E W, *The Generation of Electricity by Wind Power*, London: Spon, 1976.
9. Carter, J, *Testing an Elektro-Gemini Wind System*, Windpower Digest, Sept 76 **1**, 6, Indiana: Jester Press.
10. Sorensen, B, *Direct and Indirect Economics of Wind Energy Systems Relative to Fuel Based Systems*, BHRA Conference 'Wind Energy Systems' Cambridge, Sept. 1976.
11. Steadman, P, *Energy, Environment and Building,* Cambridge University Press, 1975.
12. Frost, L N, *Wind Power in Ireland*, Technology Ireland, Dec. 74.

DISCUSSION

Dr. G. S. Saluja (Robert Gordon's Institute of Technology)

Dr Todd described one low energy house with walls and roof insulated by 450 mm glass fibre mat. Can the use of such a heavily insulated house be justified, especially if the amount of energy required to produce the insulation is taken into consideration. I fail to understand how a ventilation rate of only 0.25 air change per hour can be justified. This level would fail to satisfy building regulations, and the house would be stuffy. Is the house in question occupied and what are the reactions of the occupants?

Dr. R. W. Todd

Because of their long life the energy content of insulation materials compared to that saved by the insulation in reduced heating over the life of the building is quite small, even at the higher levels of insulation described. The level of insulation in that particular building is perhaps a little excessive and 300 mm would have been a more reasonable figure. Our interest has been to examine the complete package – how much energy could be saved, and the price. The costs of energy in building and insulation materials are of course included. The 0.25 air change per hour is 25% of the total volume of the house and is based on an average number of occupants, providing a reasonable volume of air, rather than setting an air change rate – this is an acceptable way of meeting ventilation requirements and as the air in the house is continuously stirred by the warm air heating system, the whole volume of the house is available even if all the occupants are in one room.

J. Peach (Chartered Institution of Building Services)

I too was a little worried about the 0.25 air change per hour. Recently the work of British Gas has, I believe, shown that in a conventional type of house with normal weather stripping the order of air change rates are 0.5 to 0.75. There is a recent British Standard Code of Practice which for small gas fired heaters has made allowance for this order of air change and exempted installations which are smaller than a certain rating from having ventilation grilles fitted to supply extra air needed for combustion.

In the life time of any house it could never be guaranteed that a top up direct fired heater would never be installed which gives cause for concern at these very low air change rates.

What are the occupants reactions to such low rate of air change from the point of view of odour and are there any longer term effects on the health of occupants?

Dr. R. W. Todd

The Machynlleth house is rather different in that it was purpose built and it was possible to obtain such low ventilation rates only because of that – it would have been different perhaps had we been converting an old building. There is only one entrance so through draughts are excluded and the windows are actually sealed so there is no draught stripping problem. In an attempt to sidestep odour problems, air is extracted from the kitchen and the bulk of the ventilation for the house consists of this extract air. It is extracted mechanically through the roof using a 20 W fan. The toilets have separate extract ventilation using a timed fan operated by the door. The house has been fully occupied by a family of four for only three months. They seemed to be generally quite happy with it but that occupation was in warmer weather when the door was left open part of the day. Of course if the ventilation rate proved to be a problem, it would not be a difficult matter to increase it and incorporate heat recovery.

D. R. Wilson (University of Manchester)

At the house at Macclesfield described in Chapter 9 there is a calculated 0.3 air change per hour occurring naturally through leaks in the structure and through the window and door seals, and 0.7 air change per hour through the mechanical ventilation system. During the winter kitchen smells tended to build up despite the window located on the lee side of the room to assist natural extraction.

P. L. Johnson (Property Services Agency)

My main interest is office buildings where priorities are different from houses. The sun delivers heat and light and human beings appreciate both. In the past architects perhaps concentrated more in capturing the light than they have the heat but there seems to be a schizophrenic approach by current energy-conscious designers. Houses have been designed with completely glazed south walls and houses like the Machynlleth house with very small windows. Is there as yet an accepted philosophy of window design for the coming ages? Can you suggest what it should be – are the windows to be regarded as heroes or villains?

Dr. R. W. Todd

The solar heated building I designed at Machynlleth is used upstairs as an office and a fair amount of daylight was wanted because, producing our own electricity, we do not like using artificial light too much. The window area included is greater than would be expected in a well insulated building. A certain amount of the area is on the south side, set into the solar collector. The same glazing is used but a translucent insulation material, "Okalux", manufactured in Germany, is used in a gap in the absorber. This material is available up to about 50 mm thick and has similar insulation properties to cork (about half as good as polystyrene) and its light transmission is up to about 70%, depending on thickness. It enables very good natural lighting in a building, without the major heat loss windows normally impose.

D. R. Wilson

It is important to make a window which is variable. In some other countries windows are available with an insulated slatted shutter built in and controlled from inside. It is possible in this way to control solar gain and heat loss to suit the season and the weather. Until windows are designed first as apparatus for the control of gain and loss of light and heat, and architectural characteristics are developed out of this, no progress will be made.

K. G. Peters (British Gas)

Part of the work done by British Gas was concerned with the fortuitous ventilation obtained through crackage in conventional structures. In the type of house described by Dr Todd there will, of course, be no such crackage and ventilation, and the published work must be used with caution.

R. Burton (Taylor Young & Partners)

It seems that the high levels of thermal insulation provided in the Wates house reduce the obvious possibilities of gaining energy directly through the fabric, by using the external skin of the building to modify and take advantage of the external environment.

In particular the changing balance of solar gains and heat losses through glazing would tend to shorten the heating season and window areas could be increased to take advantage of this.

Indeed south-facing single glazing curtained at night imposes no additional heating load and on this basis the logic of increasing the capital cost to provide quadruple glazing in this situation is questionable.

Any tendency to overheating in summer due to increased glazed areas might be controlled by providing shading, increasing the thermal mass, or increasing the natural ventilation.

However, the multiple glazing of the Wates house reduces the possibility of taking advantage of free cooling during the summer months, where opening lights with adequate perimeter sealing would permit reduced ventilation rates during the heating season while taking advantage of natural ventilation when required.

Dr. R. W. Todd

By having such a low specific loss that particular house only needs heating during the worst of the winter because it makes such good use of incidental gains. Having large single glazed windows gives a nett heat loss in the worst part of the year when the heat is required most. Obviously it is desirable to be able to use free solar gain but an easy way of making use of the solar gains when they are useful while avoiding the problems of high heat loss in the winter and overheating in the summer does not seem to be available. The house already has quite a considerable thermal mass and copes very well with sunny summer weather. I am certain big south facing windows would present difficulties. It is not just a question of time constant; the steady state equilibrium temperature would be far too high in summer. In the hot summer of 1976 it was not necessary to actively cool the house and it was quite comfortable without being able to open windows. For much of the day it was cooler inside than outside. There is one openable window in the house and both this and the door could be left open.

Walking into the house on a hot day gives the same feeling as walking into a massive old church where the temperature of the air and surroundings is noticeably cooler. It is a reasonable initial impression that living in a house with sealed windows would be like living in an oven in the summer. It has not worked out like that.

D. R. Wilson

Inherent in this discussion is a vitally important point. Throughout the history of building people have taken what had to be done and made it delightful. There is no reason why this cannot be done with a window.

I am also very influenced by a paper by Wilberforce on solar gain and loss through windows on various elevations of a building. Through computer simulation he has established that for a single glazed window on a south elevation there is a nett gain during the hours of daylight eight months of the year. This goes against the interest of the glass manufacturer because if you have a project with a limited budget the answer is to install single glazing only and provide shutters.

J. Owen-Lewis (University College, Dublin)

I have a slight anxiety about the situation that would exist if a fire occurred in the Machynlleth house. Would there be an unusually rapid build-up of heat and would people using the upper floor experience difficulty in escaping, bearing in mind the single exit and the sealed windows?

I welcome the remarks made by Mr Wilson about the idea of the window as a dynamic element in the building. It is an aspect we are especially interested in in the Dublin School of Architecture's Energy Research Group, where the design of building fabric for solar gain is being studied. This seems to be an idea that should be associated with the very great emphasis that almost all contributors have given to conservation as a first priority. Even before implementing conservation measures the building must be designed to minimise demand through passive collection and through sensible measures (like correct orientation) which can be infinitely cost effective in that no additional expenditure is required.

Dr. R. W. Todd

I cannot answer the question about fire in the Machynlleth house. Because the house is open to the public, an alternative means of escape has been provided by installing a second door upstairs. This provides an opening to jump from in an emergency.

P. R. Elderfield (Building and Social Housing Foundation)

As a simple layman, I am dubious about advanced, seemingly complicated, high capital cost, alternative technologies, which will keep our homes warm and possibly comfortable. I suspect my unsophisticated parents, living in a small cottage in Coverdale on the North Yorkshire Moors, had a more practical answer to insulation and heat conservation. My father lived happily for 90 years, in our home, which had a single open fire. At night he had a warming pan and plenty of bedclothes. My parents slept with the windows open all their lives and brought me up to do the same. During the cold weather all the family insulated themselves by wearing warm clothes, and, in hot weather, we wore fewer clothes – a simple, cheap way of insulating ourselves against the elements. It does not appeal to me to live in a very expensive boiler house, together with 600 gallons of water, with one small sealed window and one entrance door. I believed that alternative technologies were concerning themselves with a simpler way of life, and less expensive capital resources. I despair of this pathological concern with warmth and comfort. This never had anything to do with creating the Great in Great Britain. I question the whole philosophy, the whole set of values which we seem to be establishing.

Dr. R. W. Todd

I agree with sensible clothing and sensible temperature. Personally, and at our Centre, we live with lower temperatures and wear clothing appropriate to the weather. Nevertheless, if alternative technology is to be taken seriously it is necessary to be able to argue that it can provide a certain comfort level. Slightly different comfort levels have been mentioned by different speakers arguing for example that lower air temperatures and higher radiant temperatures are the best solution. If alternative technology is to be made more than just a minority interest, it is essential to be able to show it is capable of providing reasonable comfort levels even if at the same time the protagonists recognise it is much healthier to live at a lower temperature.

I do not think we can retrace our steps towards the houses and lifestyles of the past. In a few years it will not be possible to just go out and get wood or whatever other solid fuel was traditionally burnt. If we have to move away from traditional fuels towards ambient energy sources we must learn to live using considerably less energy and with much higher efficiencies than the open fire can generally provide. Mr Elderfield described the solutions put forward as very expensive. At the moment they are, but, there are two mitigating factors. One is that the whole field is as yet very new and the present prices of one-off items are much higher than the prices of similar items if mass produced. Efficiencies will also improve as will the ability to live with the equipment without the house looking like a ship's boiler room. The second factor comes from considering and assessing the real costs of the future based on present technology, which seems to depend on coal and massive use of nuclear power. A massive nuclear power development, to satisfy energy demand will depend on electricity generation or perhaps hydrogen, both produced at low efficiency, with collasal waste heat problems. The actual cost per usable kilowatt is astronomical. I think projected cost estimates of the nuclear type future have not realistically accounted for the enormous problems of transmitting a large proportion of our energy demand as electricity or hydrogen, and dealing with the vast quantities of waste heat and radioactive waste involved.

Dr. J. C. McVeigh (Brighton Polytechnic)

I agree entirely that the cost of a nuclear future is escalating rapidly beyond any possible benefit. We must look forward to a future which contains the energy mix illustrated in Dr Todd's paper. On the possibility of cost reductions, it is interesting to note that when the ballpoint pen first appeared on the market 25 years ago the purchase price was £5. It is now 5p. This level of cost reduction could be anticipated with direct electricity from solar cells. The solar roof which could provide electricity directly is a distinct possibility. It is inevitable that the initial stages of any new developments, such as wind power with modern windmills, are expensive because it is necessary to support the people who are doing the work without selling any product. This work needs to be done now. It will be too late in 25 years time because North Sea oil and gas which will provide energy support during the development of alternatives will by then have been depleted.

D. R. Wilson

I am sympathetic to the desire to use simple solutions. One of the snags is the tendency to generalise about solutions by statements such as "what was good enough for our parents is good enough for us" or "we must have 19°C throughout the house". We seem to have forgotten how to take opportunities for change, variation, modulation of experience, etc, all of which are important. In the long run the whole concept of alternative technology will only be accepted by people if it also engenders delight.

8. A Wind/Solar Project in Local Authority Housing

J. C. McVeigh and W. W. Pontin

INTRODUCTION

There has been a very rapid increase in the number of energy-conscious housing schemes throughout the world in the past few years. The scale of activity has ranged from individual houses to large district heating schemes. Some of these applications have involved solar heating and/or heat pump systems and much of this work has been reviewed by Shurcliff (1). In the United Kingdom the earliest example of a house designed to use solar energy combined with a heat pump for space heating was built in 1956 by Curtis (2, 3). This was followed by the most widely publicised solar building in Europe, the annexe to St. George's School, Wallasey (4), where it has been shown that at least 30% of the total heating requirements are provided by the passive collection of solar energy through its large south-facing double glazed solar wall. In February 1977 the Government announced a £3.6 million four-year programme of research, development and demonstration of solar energy in which a major feature would be solar water and space heating. Already there are probably at least twenty houses or buildings with a substantial solar area being monitored, although some of these have not been disclosed for commercial reasons.

In parallel with work on the direct use of solar energy for heating, many countries now have wind power research programmes, and designs ranging from 100 watts to 2 or 3 megawatts are perfectly practical by today's standards. As yet, there is no official UK wind energy programme, but a very substantial amount of development work has been carried out by the Wind Energy Supply Company (WESCO) with support from the NRDC. As part of the WESCO policy of seeking out new applications for wind energy, the integration of a bank of fixed ducted windmills to provide part of the electricity demand for a local authority housing scheme has been considered. The local authority, the Lewes District Council, has provisionally allocated a level site at Peacehaven, Sussex, about 1 km from the sea and approximately 50 m above sea level. The total site area is 0.466 hectares of which the experimental low energy housing group has 0.268 hectares. The main features of the proposals are as follows:—

- (i) The use of the complementary nature of wind and solar energy.
- (ii) The energy attack is towards oil and electricity. The houses have normal electricity and gas services and no restrictions are placed on the use of smokeless fuels.
- (iii) All housing units will be provided with insulation substantially greater than the 1975 Building Regulations minimum to give an overall U-value of 0.6 $W/m^2 {}^\circ C$.
- (iv) Each housing unit has hot water piped to it from a hot water main. This hot water supply is metered.
- (v) The hot water main is supplied from a common energy supply and storage system.
- (vi) Heat pumps, either at individual or group level will also be included.
- (vii) Space heating will be based on the concept of large area comparatively low temperature units.
- (viii) Full liaison would be established and maintained with electricity and gas utilities on the technical and economic aspects of the scheme.

AVAILABILITY OF AMBIENT ENERGY AT PEACEHAVEN

There are no direct or reliable long term statistics from the site, but various readings of wind speed which have been taken over the past three years indicate that 6.0 m/s is a reasonable mean velocity and this has been used in all calculations. A more detailed treatment of the wind regime is given in Appendices 1 and 2. For solar radiation data, the figures derived by the BRE (5) from Kew data and applied to a slope of 30° to the horizontal have been used, and the monthly and annual totals for several other slopes are given in Appendix 3.

For design purposes, both wind energy flux density and incident solar energy were expressed in W/m^2 and are given in Table 1.

This illustrates the complementary nature of wind and solar energy, but does not take any account of conversion efficiencies. This has been done in Fig 1, where the wind energy figures have been multiplied by 0.3 for conversion into electricity, and the solar energy

Figure 1 **Effective energy flux density – monthly basis.**

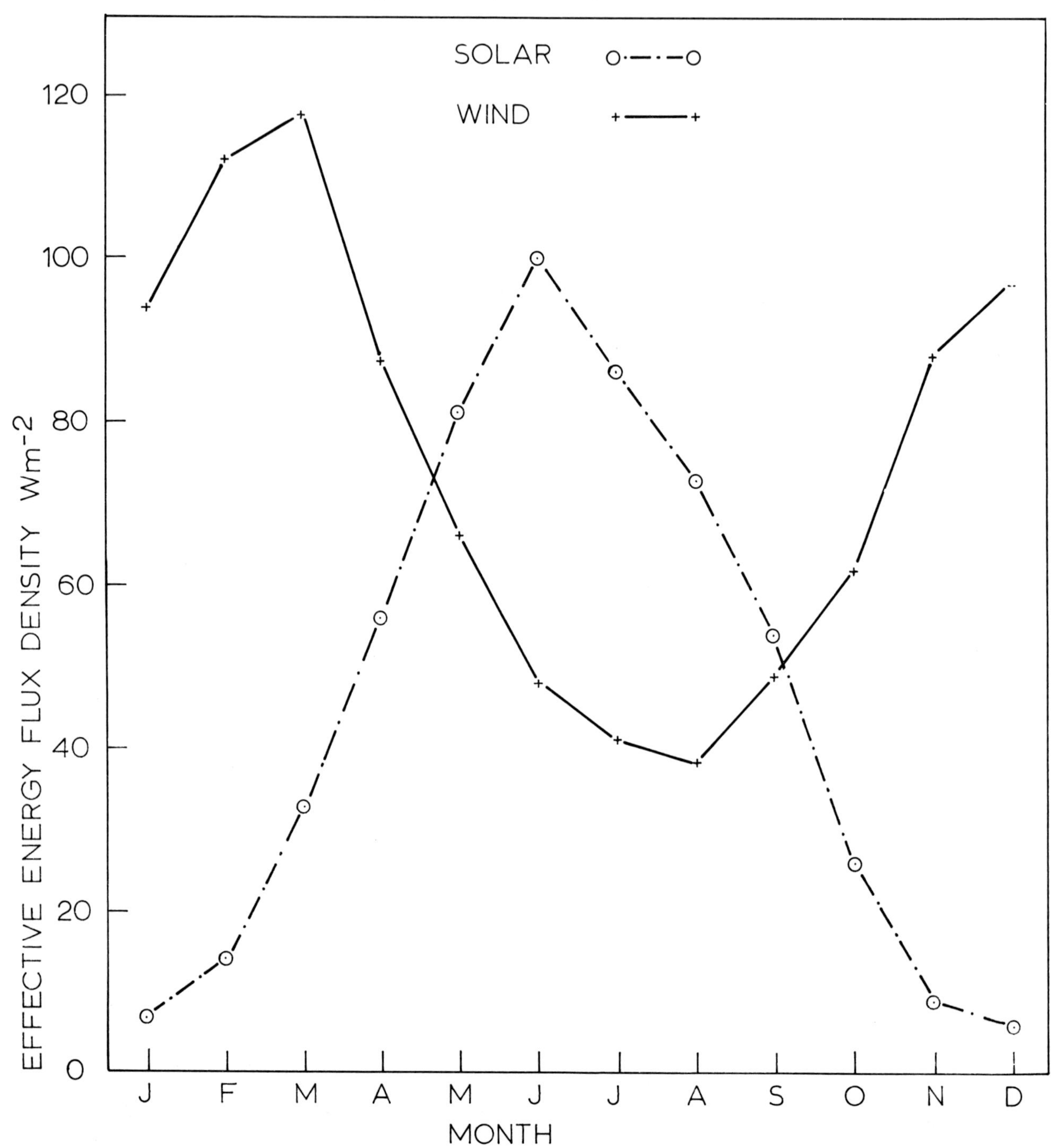

figures by an appropriate factor for the month. The efficiency of solar collection depends very much on the intensity of radiation. For June and July the efficiency can be assumed to be 50%, but by October it would drop to 30% and for the next three mid-winter months to 20%. It can be seen that a combination of solar energy to provide the main heat supply to appropriate storage, coupled with wind energy to give the additional mid-winter boost has very considerable potential.

Table 1

Month	*Wind Energy Flux Density W/m^2 (vertical)*	*Solar Energy W/m^2 (at 30° to horizontal)*
January	312	34
February	373	56
March	392	110
April	290	139
May	220	179
June	160	199
July	137	177
August	128	162
September	164	135
October	208	88
November	293	46
December	308	32
Year (approx)	248	113

Basic design parameters

Housing Units. The basic types have been included in the scheme and the main features are given in Table 2. The total floor area of 586 m^2 is used for the calculations. Insulation to a standard approaching 0.6 $W/m^2 {}^\circ C$ will be provided, so that an annual demand of 150 000 kWh for space and water heating can be assumed. The units will be grouped in three blocks, forming the sides of an open U.

Table 2

Type	*No of housing units*	*No of persons*	*Net floor area per unit (m^2)*	*Total net floor area*
Two storey block containing four flats, 2 person, 1 bedroom	4	8	46	184
Single storey block containing three flats, 2 person, 1 bed-room	6	12	46.86	281.16
Two storey block containing two houses, 3 person, 2 bedroom	2	6	60	120

Solar Collectors. The minimum area of collector needed to supply more than about 40% of annual demand in the UK is 0.25 of the heated floor area. As only standard housing units will be considered, the collectors will be placed close to the houses, forming part of the artificial slope behind a proposed garage block. A total area of 200 m^2 of collector will give a ratio of collector/floor area of 0.34. A unit cost of about £45/m^2 delivered, but not installed, can be taken in April 1977. The collector type will be chosen more for the ability to stand up to the climatic conditions than performance, as an analysis published elsewhere (6) shows that the performance characteristics of the great majority of collectors are very similar under good radiation conditions and low temperature rises above ambient.

Figure 2 **Preliminary circuit design.**

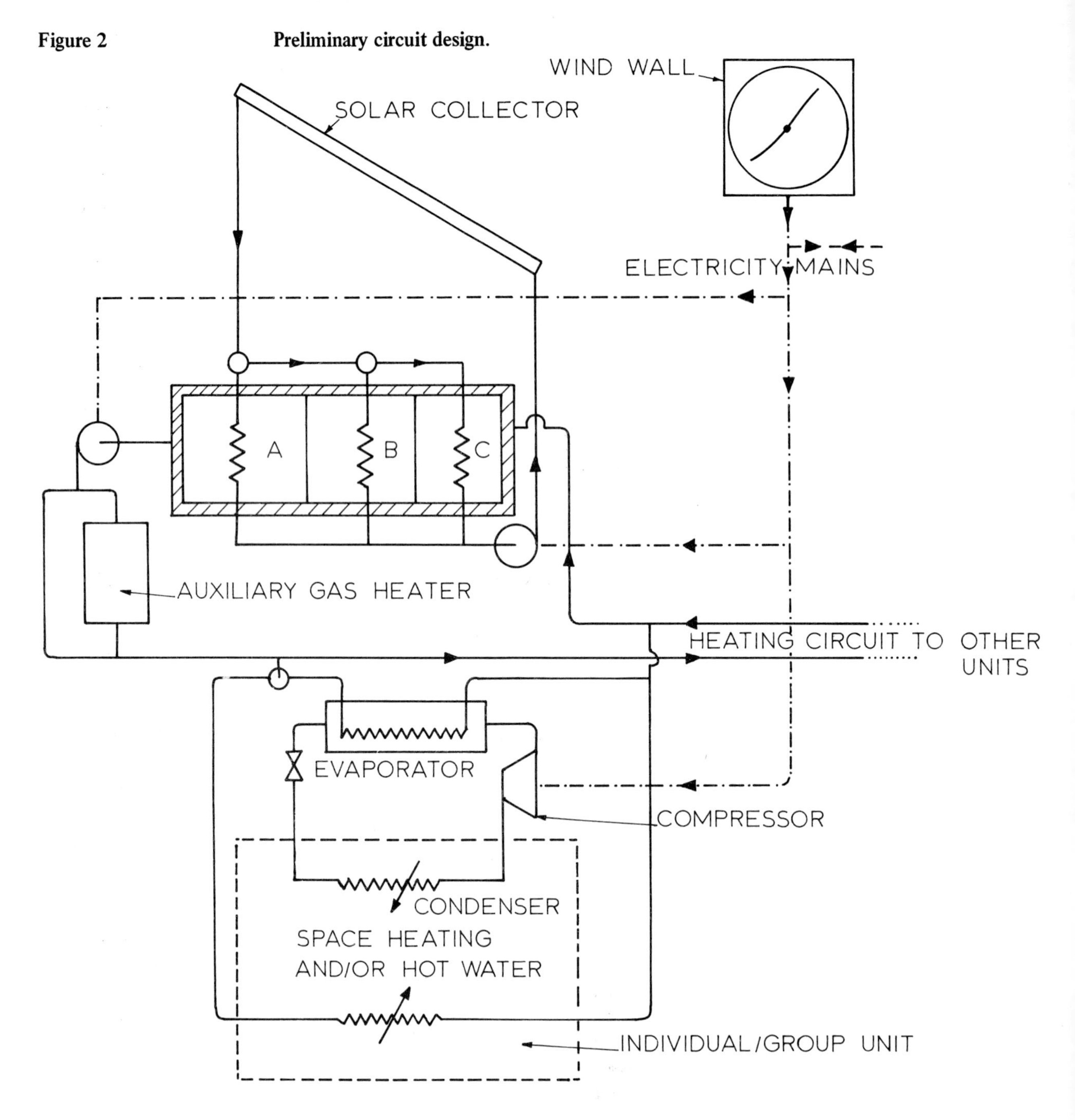

Wind Wall. The vital characteristic of wind is that it can produce electricity at mains frequency, therefore serious consideration has been given to the indirect use of wind energy to reduce the electricity bills to the householders. Discussions have taken place to establish the possibilities of directly connecting the mains frequency producing windmill units to the electricity supply network. The problem appears to be economic, rather than technical, in that the Electricity Board could not forego the capital portion of their tariffs.

One of the major objections to a conventional windmill in an urban situation is the visual impact. Ducted windmills, fixed in position and direction, can overcome this problem. The concept of a series of ducted windmills, each based on a 2 m cube, has been proposed for Peacehaven. The venturi-shaped entrance and exit sections compensate to a considerable extent for the relatively small periods when the wind is unable to pass through the circular central section which contains the bi-directional blades. It is proposed to place the windmill at the top of the south-south-west facing solar panels. The northerly side could conventionally be used as a single storey garage block with long term water storage underneath. Detailed production costs are not yet available for these machines. Twenty

Figure 3 **Mean daily energy – monthly basis.**

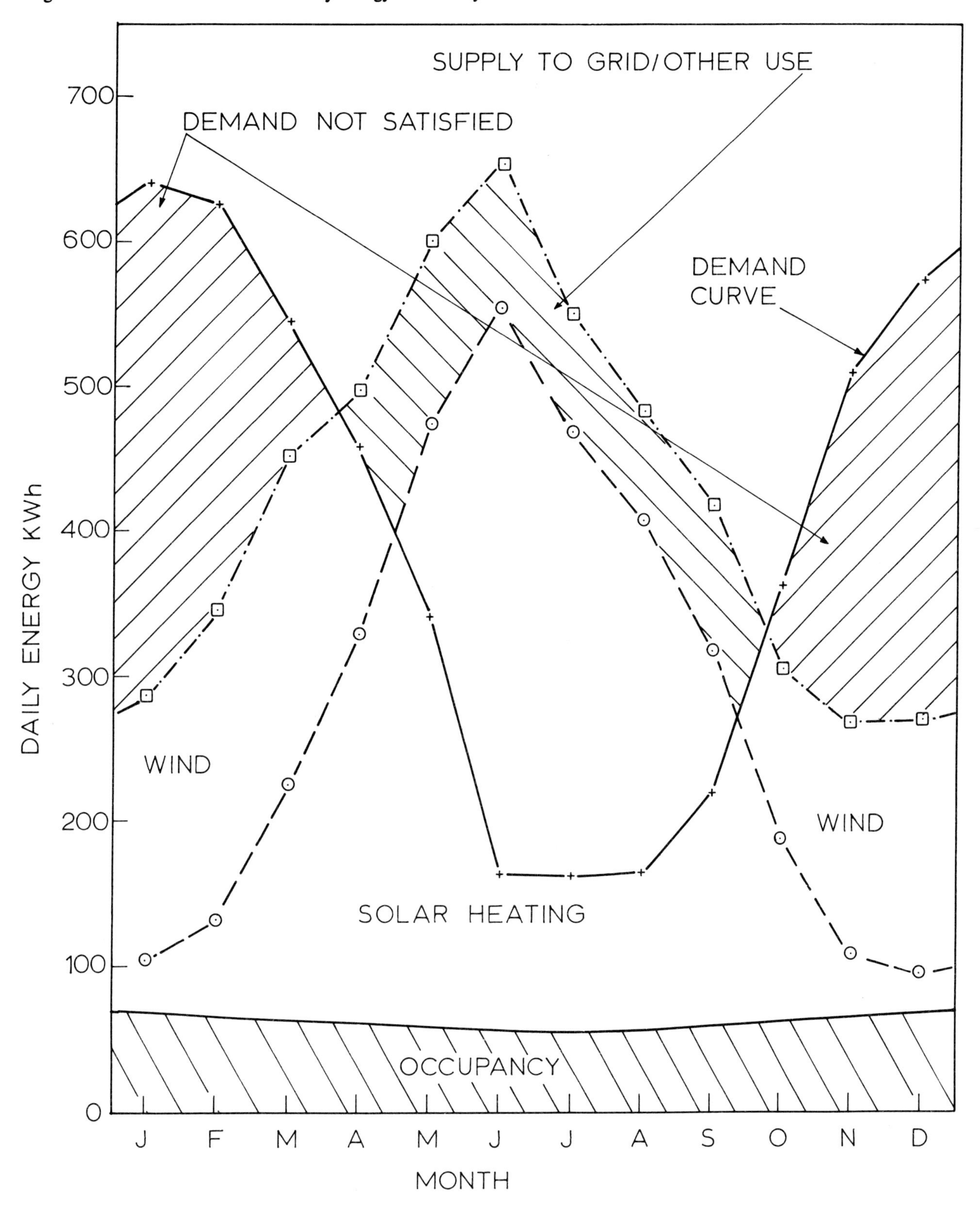

units, with a total area of 80 m^2 are suggested.

Storage System. An examination of the characteristics of many solar space heating systems shows that for a well insulated house in typical UK situations a ratio of water storage volume (litres) to heated floor area (m^2) of 150 is a very reasonable compromise to aim for (6). This can provide about 60% of the total demand for the collector area/heated floor area ratio of 0.34 chosen above. Doubling the storage volume will improve the annual performance, but only by a factor of 1.1, so a storage volume of 10 m^3 has been chosen. This gives a ratio of 170. An important feature of the storage system will be the internal division into a large, medium and small volume – an approach suggested and adopted at the National Centre for Alternative Technology (7). This enables considerable use to be made of solar radiation in the late autumn, when conditions may be appropriate for collection at good efficiency at 35°C but not at 55°C. The initial use of the heat in the smallest storage section could drop its temperature well below the level of the rest of the system and efficient solar collection could take place.

Preliminary Circuit Design. The main features of a possible circuit are shown in Fig 2. The collectors are shown with heat exchangers on a closed circuit, but could equally well be direct, with no heat exchangers. A heat pump is shown in the individual or group units, but consideration is also being given to one central unit. The concept of a dual source heat pump and hot water supply system, using ambient air as the alternative source, will also be considered (8). The mains electricity supply can also be fed from the wind wall under appropriate conditions.

Energy Demand and Supply Pattern. The gross space heating and domestic hot water demand of 150 000 kWh per annum is shown as a daily demand, averaged for each month, in Fig 3. The contribution of solar heating and wind combined can satisfy just over 100 000 kWh or 66.8% of this demand and occupancy gains account for a further 6.7% giving a total of 75.5%. Direct solar gain to the houses has been ignored in these calculations. There is a net gain during the summer months of electricity supplied for use in the house, or directly to the grid, of about 15 000 kWh. This should be regarded as a positive gain to the householder, further reducing the total annual bills paid for combined heating and electricity, and amounting to some 10% of the total space and water heating demand.

SUMMARY

The work described above has attracted very considerable interest, both at local and national level. Detailed financial estimates are being prepared and support is being sought both for the additional capital required for the solar/wind system and for continuous monitoring after installation. The combined solar/wind approach has been shown to be appropriate for conditions at this site, where there is a relatively good solar climate and a moderate wind regime.

ACKNOWLEDGEMENT

The authors would like to thank the original proposer of the scheme, Michael J Blee, MArch (MIT) RIBA, FRSA, FRGS, MSIAD, of the Michael Blee Whittaker Partnership for establishing the initial dialogue between those concerned in the development of these proposals and with the Lewes District Council. Michael Blee Whittaker Partnership are members of the Environmental Design Group practice based in Lewes, Sussex.

REFERENCES

1. Shurcliffe, W A, *Solar heated buildings – a brief survey*, 13th Edition, Cambridge, Mass, January 1977.
2. Curtis, E W J, and Komedere, M, *The heat pump*, Architectural Design, June 1956.
3. Curtis, E W J, *Solar energy applications in architecture*, Department of Environmental Design, Polytechnic of North London, February 1974.
4. Davies, M G, *Model studies of St George's School, Wallasey*, JIHVE **39**, 77, July 1971.
5. Courtney, R G, *An appraisal of solar water heating in the UK*, Building Research Establishment CP7/76 (1976).
6. McVeigh, J C, *Sun Power : an introduction to the applications of solar energy*, Oxford: Pergamon Press, 1977.
7. National Centre for Alternative Technology, Machynlleth, Powys, Wales.
8. Sakai, I, et al, *Solar space heating and cooling with bi-heat source heat pump and hot water supply system*, Solar Energy **18**, 6, 525-532, 1977.

APPENDIX I

Treatment of Wind Energy Regime

Traditionally, the velocity/duration curve, expressed on an annual basis, has been the main method of treatment of wind regimes (windmill division). The introduction of the computer has reduced the laborious computations involved to trivial proportions. In recent years, the Wiebull distribution has been commended by many authors for wind regime methodology and the experience of the present authors bears this out.

The probability that a wind speed greater than V (m/s) exists (at a standard height of 10 m) may be expressed as

$$P(>V)$$

where $O \leqslant P(>V) \leqslant 1$.

The Wiebull Distribution assumes

$$P(>V) = \epsilon^{-\left[\frac{V}{C1}\right]^{K1}}$$

where ϵ = 2.7801
C1 = Wiebull constant having units of m/s
K1 = Wiebull exponent

(Note for computer use the suffix '1' eg C1 is annual, suffix '2' is monthly, suffix '3' is daily, etc.)

Properties of C1 and K1

Normally C1 is some 10-12% greater than the mean annual wind speed whilst $0.9 < K1 < 2.0$. Note also the following subsidiary properties:

(a) Velocity/Frequency Distribution

$$P(V) = \frac{K1}{V} \cdot \left[\frac{V}{C1}\right]^{K1} \cdot \epsilon^{-\left[\frac{V}{C1}\right]^{K1}}$$

(b) Mean Wind Speed

$$V = C1 \cdot \Gamma\left(1 + \frac{1}{K1}\right)$$

where $\Gamma(x)$ is the gamma function = $(x-1)!$

(c) Median Wind Speed

$$V_{med} = C1 \cdot (Ln2)^{\frac{1}{K1}}$$

(d) Modal Wind Speed

$$V_{mode} = C1\ (K1-1)^{\frac{1}{K1}}$$

for $K1 > 1$

Wind Energy Flux Density

WEFD = Kinetic energy in the wind passing through unit vertical cross section in unit time.

WEFD is expressed in W/m^2 and varies between a low of about 20 and high of about 1500 W/m^2.

$$WEFD = 0.5 . p . C1^3 . \left(1 + \frac{3}{K1}\right) W/m^2$$

where p = density of air (1.226 kg/m^3 at sea level).

Energy Factor. Because the cube of any series of numbers (not all equal) is greater than the cube of their mean, any wind energy analysis based on assuming that the mean annual wind speed blows continuously is totally wrong.

$$\text{The Energy Factor} = \frac{\text{True WEFD}}{\text{mean WEFD computed from the mean wind speed}}$$

$$EF = \frac{\Gamma\left(1 + \frac{3}{K1}\right)}{\Gamma\left(1 + \frac{1}{K1}\right)^3}$$

The Energy Factor can vary from about 1.5 to as much as 8 in exceptional cases. In the UK a 'typical' figure would be about 2.1.

The Peacehaven Scheme

There are no long term, reliable statistics for the site. The following represent a careful assessment.

Annual

Wiebull constant C1	=	6.76
Wiebull exponent K1	=	1.91
Mean wind speed (m/s)	=	6.00
Median wind speed (m/s)	=	5.58
Modal wind speed (m/s)	=	4.59
WEFD (W/m^2)	=	264.7
Energy Factor	=	2.00

APPENDIX II

Treatment of Boundary Layer Profile

Architectural features have not yet been settled but boundary layer effects, specifically variation of wind speed with height, could be treated with a power law profile.

Typical figures might be:–

Wind speed at top of friction layer	=	50 m/s
Height of friction layer	=	350 m
Power index, a	=	0.18

Giving

$$\frac{V_H}{V_{10}} = \left[\frac{H}{10}\right]^{0.18}$$

where V_H is the wind velocity at height H.

APPENDIX III

Monthly and annual totals (MJ) of solar energy incident on 1 m^2 of collector at different angles to the horizontal, derived by BRE computer programme from Kew average solar radiation data for 1959-1968. (After R G Courtney, *An appraisal of solar water heating in the UK*, CP 7/76, Building Research Establishment, January 1976).

MONTH	*TOTAL*			
	30°	*45°*	*60°*	*90°*
January	90	105	105	100
February	135	145	140	125
March	295	310	295	240
April	360	345	325	235
May	480	455	415	285
June	515	475	425	280
July	475	450	400	265
August	435	420	380	275
September	350	350	335	265
October	235	250	245	210
November	120	130	135	125
December	85	95	100	95
Annual	3575	3530	3300	2500

DISCUSSION

Dr. R. Critoph (Energy Research Group, Open University)

I would like to ask Dr McVeigh to explain the aims and objectives of the project. Is it to act as a demonstration, is it a finished product that can be duplicated elsewhere, or is its purpose simply to try out some interesting ideas? Secondly, as the houses are purpose-built, why not use higher standards of insulation. This would reduce the store size, the solar panel area and also, incidentally, the amount of money needed to build it in the first place. Also, has the danger of stones or other projectiles being thrown through ground level solar panels or into the windmill been considered? When panels are mounted on a roof they are almost immune to that kind of damage.

Another point regarding the windmills is that if the windmills are presented to the wind as described, would it not be as good and less expensive to have a vertical axis windmill on its side in the rectangular area rather than a number of discreet and probably very expensive machines?

Dr. J. C. McVeigh

A vertical axis windmill on its side has not been considered. The aim was for a modular approach suited to any project by installing the appropriate number of standard modules. Another aim was to look at the integration of wind and solar energy in a local authority housing project. At this stage we cannot predict how many wind units or solar collectors will be included in the scheme – it may not be exactly as presented in the paper. We are hoping the insulation values will be much higher. Many local authorities are now appreciating that by spending more in the initial stages they can save on annual running costs.

The problem of small boys and stones is always with us – there are transparent materials that can withstand stones and it is most likely the solar panels will not be covered by glass. It might be necessary to sacrifice the efficiency of the wind wall by protecting the ducts from stones using wire mesh.

J. Peach (Chartered Institution of Building Services)

Dr McVeigh suggested in connection with the wind generator scheme he described that at times when there is surplus wind power it could be connected to the mains. It sounds simple but would there not be problems with synchronization and load control? Would those operations be done automatically, and if so would such automatic controls be cost effective bearing in mind the relatively small contributions to the national grid system?

Dr. J. C. McVeigh

It is hoped to feed electricity into the grid primarily to demonstrate the feasibility and that there are no technical problems that cannot be overcome quickly and simply. It will be cost effective particularly in the smaller communities, such as on some of the Islands of Scotland. At present, it could be argued that wind development work is not cost effective, because we are still in the pre-production prototype development and test stage. A true "mass-production" price of a windmill will be well below current prices. Other countries do have national programmes for feeding windmill generated electricity into the grid.

D. W. Dallinger (Troup, Bywaters & Anders)

In the courtyard arrangement described, the houses seem to block the wind on to the wind wall – have you looked at the turbulence caused by this and would it be better to have a V-shape on to the wind wall to concentrate the wind energy on to it? I read recently of a district heating scheme in Japan using solar energy that had been discontinued because it was not cost effective. Would Dr McVeigh like to comment on cost effectiveness?

Dr. J. C. McVeigh

It is essential in assessing cost effectiveness of all solar energy and wind energy systems that payback in five years must not be expected. It is important to consider too the benefit accruing to the community and to the country as a whole whenever solar heating and wind systems are installed. Every system will save some energy and reduce the importation of oil or the use of indigenous fuel resources. Many people lost more in the depreciation of their cars in the last year than it would have cost to install a solar water heating system. Assessed in this way solar heating begins to look rather more attractive.

I agree that it seems there is some blocking of the wind, but the building is single storey with quite a large gap over gardens before the wind reaches the wind wall. It would be quite possible to have a V-shaped avenue of trees at the back of the wind wall. These experiments will not show any financial profit, especially as many of the people involved are giving their time on a voluntary basis because they believe in the future of wind power. It is hoped that the Government will soon put forward a wind energy development programme to match the recently announced £3.6 million solar programme.

K. G. Peters (British Gas)

I suggest there would be considerable turbulence at ground level, especially with the houses to windward. I wonder if allowance has been made for this, or is it yet to be established in the wind-tunnel tests?

The shape of the entry to the windmills was, I understand, to collect wind from varying directions as far as possible, but was it also to obtain a Venturi effect through the blades? If so, this would be lost with the wind coming from other directions.

I am interested in the type of water-to-water heat exchangers to be used – whether they will be the high efficiency plate type or some other design. There can be problems of fouling from scum and lint when recovering heat from waste water, even from domestic bathing and washing, and provision must be made for cleaning the heat exchanger. I have experienced this problem when attempting to recover heat from the waste water of launderettes, when scumming and the difficulty of cleaning was a serious obstacle.

Dr. J. C. McVeigh

Regarding turbulence, the wall may not be straight but slightly curved. It has been shown that the performance of a wind turbine or a windmill can be enhanced by a factor of up to 5 by having an appropriate shape in front of it shedding vortices. We hope to get a benefit from having the sloping collectors going up to the wind wall. It may be possible to achieve a similar effect on the other side using the slope of the garage roof. Calculations indicate that about 77% of the energy available from a conventional windmill could be collected by a wind wall. There is a Venturi effect which increases the velocity through the centre of the wind wall. The cube law effect helps to overcome some of the disadvantages of non-operation with the winds at right angles. Until wind-tunnel tests have been carried out the calculations cannot be confirmed. Use of straightforward tubular heat exchangers is planned, with a closed circuit.

Dr. R. W. Todd (National Centre for Alternative Technology)

The fouling question particularly applies when heat is being recovered, for example, from sink waste. We have done what we can to avoid trouble and we do not have enough experience to know how well it will work. An important point about heat exchangers in solar heating, or indeed heat pump systems, is

Figure A 18 metre diameter windmill capable of powering 30 low energy houses.

the possibility of contamination risks to domestic hot water supplies. The Water Authorities will not accept a heat exchanger carrying toxic antifreeze or refrigerants in a domestic hot water tank unless the heat exchanger has no joints whatever inside the tank. The only acceptable design is a tube type heat exchanger which passes through the skin of the tank, coils round inside, and out again. The only joints in the pipe are outside the tank.

D. A. Burnett (Burnett Pollock Associates)

I would suggest one possible approach to the use of wind in the immediate vicinity of buildings is to make use of the wind turbulence. On very tall buildings structural engineers often advise the incorporation of wind slots through the building to provide a release of wind pressure. This causes a fairly concentrated high speed wind current to pass through. There seems to be scope for making use of this energy. On the visual aspects of windmills, it strikes me that we may be talking about something that is not a problem at all. The Plain of Lasithi in Crete has 1000 windmills and they are the principal tourist attraction of the area.

Dr. J. C. McVeigh

We hope that wind wall modules can be integrated into buildings. The building would have to be designed to withstand the horizontal thrust and the problem of noise would also need careful attention.

Dr. R. W. Todd

Even with a massive wind power programme producing a large fraction of our energy supply, we would only require one quite small windmill per km^2 (200 acres) if most of the machines were sited on land. The separation would be much greater if we used larger machines or put them in clusters. If most of the big machines were positioned in the North Sea, the total visual impact, providing land based machines were of attractive proportions, would be quite acceptable. (See Fig A).

9. Experience from the 'House for the Future' Project of Granada TV (Part 1)

B. Trueman

AMBIENT ENERGY & BUILDING DESIGN

I suppose the word that informed all our thinking about the House for the Future – and I mean both the structure itself and the television series – was *relevance.*

The original idea from which the project sprang was simply to construct a primitive dwelling out of naturally available materials and allow – or compel – a television 'personality' to live in it for a week. I ruled that out on the grounds that its relevance to anyone's real needs was, to say the least, minimal.

When later we had to decide whether to build from scratch or convert an existing building it was a desire to relate to the greater proportion of the viewers that steered us towards that derelict heap in the Cheshire hills near Macclesfield. Paradoxically it was 'relevance' again that produced an energy conserving and converting house whose elaboration and cost would put it beyond the reach of most people, because, contained within that *over*-equipped structure, are many variations on the theme of energy conservation among which I hope most people will be able to find the clues to a solution relevant to their own specific needs.

Persuading Granada Television that they should go into the construction industry was not easy – but neither was it as difficult as it might have been since Granada's developing philosophy of making programmes that have a direct relationship with public needs (Reports Action, This is Your Right, etc) coincided with my proposal to make a programme that was of immediate and urgent relevance to a very broad section of our viewers.

So, on the basis of some early research, I put forward the programme idea in August 1974 and it was given the go-ahead two days before Christmas that year, by an executive who thought that the Cambridge Autarchic housing project was the work of Magnus Pyke, that our own house would be a sort of plastic dome, sprouting antennae and that transmission of programmes showing progress on the house could start three months later.

I mention these misconceptions not to pillory the higher echelons of television but to give some idea of the difficulties we all faced in reconciling the three main functions of that building. It had to be a home, it had to be a test-bed and it had to be television material. To satisfy all three requirements without betraying the expectations of the Grant family, the scientific world and the Television Company was a juggling act which called for a constant re-examination of our aims and assessment of our achievements.

We were fortunate in having a strong and happy team. Researchers, secretary, directors or producer, we found ourselves playing roles for which we had neither training nor experience. No-one had a science degree and all played a part in processing technical information so that it should be comprehensible to the average viewer without, we hoped, being condescending to those with above-average scientific knowledge.

And we were fortunate in our choice of Geoff and Lynn Grant who, with their two young daughters Helen, then three, and Caroline, six months, were to live in the house they would help to build. Chosen from 250 applicants, they were practical, articulate, involved and resolute.

The architect for the project was obviously going to be the keystone of the whole elaborate structure and it was in this area that somebody up there – other than Lord Bernstein or Eric Lyons – demonstrated an incomprehensible affection for me. I met Donald Wilson quite accidentally, took in the style of the work he'd done on his own house, looked at his bookshelves and talked to him briefly. Then I backed my intuition, took a deep breath and asked him if he'd like to design our house. And he took a deep breath and said 'Yes'.

The final basic ingredient was the property we were to transform. Two researchers combed the area the Grants had chosen, and one of them came up with the solution which was inspected by Geoff and Lynn and by Donald Wilson. They found it acceptable, I found it horrific but we bought it anyway.

And so we came to matters of construction at which point the architect, simultaneously handicapped *and* spurred on by the demands of television, came into what could only loosely be called his own. The technicalities of his solution are contained in the following chapter.

Looking back over the 2½ years since the inception of the project I am most of all surprised that we managed it at all. The team was as small as it was inexperienced in this area of building and the time available was, by any normal standards for scientific research, ludicrously short. Moreover it is not the policy of Television Companies to innovate – rather do they parasitically record other people's innovations. I think that Granada Television Limited deserves praise for a courageous investment: it was not, however, as expensive a programme as many less positive and less reflective of the viewers' real needs and I would like to see both ITV and the BBC play a practical part in helping to educate a nation that is backward in its understanding of architecture and design.

Within the general view of our experiences several peaks stand out. I shall merely observe them without trying to organise them into a tidy philosophical landscape.

The first point is that better buildings seem to me, a layman, to be a matter of common sense. Much research is being done to uncover principles used by our forbears as a matter of course. Before we learned to tap the solar battery of coal, oil and gas our houses were, of necessity, built as a conscious reflection of the environment. Most of the faults and evils of modern building seem to me to spring from pretty futile attempts to defy nature; now that the solar battery is all but flat we shall have to revert to the earlier approach. It is ironic that of all the techniques employed on the Macclesfield house, the cheapest and simplest show the greatest advantage, offer the best return on investment and promise the least problems of maintenance.

This in turn would seem to indicate that there is scope for the individual to wean himself from the breast of a beneficent but possessive Big Momma Society and do something about his own energy consumption and requirements. We need to make sure that his efforts are not penalised through increased rates – indeed, it would make sense actively to encourage him by, for instance, removing the VAT on insulation materials. What hope there is for this kind of common sense approach is a matter for doubtful conjecture when, in the area of public building, councils are not allowed to add insulation beyond the confines of the cost yardstick on the grounds that "only the tenant would benefit"!

The project team has had its eyes opened over the follies of short term accounting – of "build cheap now and to hell with the maintenance budget". I only plugged my bookshelves into the plaster once – nationally we do it all the time.

Any builder I have talked to has pointed out that he will greatly increase his chances of selling a house by sticking £100 worth of gewgaws on the outside rather than by spending £200 on better construction. The British public do seem to be among the most pretentious and the most ignorant in their housing requirements. I cannot but feel, however, that that is how the business wants them – "Prestige" always looks so much prettier on a billboard than "Solid". If that is not the case then I think it would benefit our own generation and those who follow if the public were better educated in matters of housing and I feel that the many professional bodies that relate to the Construction Industry could do more to bring that about.

My optimism that this could be achieved has been reinforced by the correspondence our programme generated. Never less than 150 letters a week came in while the programme was on the air and we have had as many as 500. Many of these were intelligent cries for help over energy conservation techniques or such recurrent problems as condensation.

Many of the people want to understand the machines they live in, many want to work on them. Ought they not to be helped?

I cannot believe that it is beyond the capabilities of the various official bodies to organise some sort of advisory service to the general public. In the long run, having a better educated consumer could only benefit the responsible majority of the construction industry and improve the all too often tarnished image of the builder and the architect.

The writers of many of those letters also encouraged us to believe that we had made perhaps more right decisions about the attitudes we took in the programmes than we had any reasonable right to expect. From the start we determined to avoid the edited perfection that is apparent in most filmed or recorded programmes because we felt that to present the sort of techniques we were trying to put into practice as being easy would only be a disservice to the public. Where we made blunders or struck obstacles we showed them; blood, sweat and embarrassment were all part of the story.

Much of it could have been avoided – let alone edited out – if we had more frequently settled for proprietary, ready-made solutions. It was, however, always our intention to encourage people to provide for their precise needs. In other words, we were saying "Don't do as *I* do – think as I *think* and then do as *you* do". Wherever it was possible to do so we demonstrated basic analysis and basic techniques of building and energy conserving needs so that people could come up with their own solutions.

I mentioned obstacles – some of them were in the form of planning or building regulations and again the viewers' letters indicated that this is not unheard of in other building ventures. The individuals administering those regulations were never less than helpful but it is clear that in many cases the legislation is designed to cope with a situation other than that of the present and that like most legislation it is not intelligently flexible enough. For a nation to show enterprise it must be free from the discouragement of too many petty, unintelligent and negative controls.

The programme then and the project as a whole have been surprisingly successful – how successful the house itself is in terms of energy conservation, how near its performance comes to the predictions outlined by Donald Wilson in Chapter 10, remains to be seen from the quantitative analysis that is being organised now. What we do know is that it is certainly working as a thermally polite, convenient, and practical home which is a growing source of delight to the Grants, both adults and children. In summer there is, of course, an abundance of "free" hot water while at the same time the internal temperatures, even in the heat of Summer '76 have been kept well within the comfort levels. In the coldest weather of the winter of '76-'77 the internal temperature was maintained at a steady 20°C by a 3.5 kW solid fuel boiler.

In early winter there were problems with the solar system when one of the simpler elements – a "frost-stat" failed to function. The resultant freeze up caused more damage to the pump and the electronic circuitry than it should through the further failure of an over-load cut-out. So much for tried and tested technology!

I had hoped to be able to report on the energy derived from the windmill. All the major elements of this have been tested – the output from the alternator has reached the rated output of 60 amps at 24 volts, we know that Dr Cartwright's feathering device works and that the wind-sensitive trigger will swing the rotors edge-on to the wind at a pre-determined wind-speed. The mill is not, however, providing any input to the house because we are anxious to make the mast which carries it more rigid and there has not been time to carry out the modifications necessary. In general our experience of the windmill would lead us to doubt the practicality of wind generators for domestic use where electricity can be taken from the grid. The promise held out by the vertical axis device designed by Dr Musgrove of Reading University could change that opinion.

It was one of the great frustrations of the building period when we were making the programme that the 'simple and inexpensive' devices for collecting and storing free energy became more and more complex and more and more expensive as they were designed to work in reality rather than in our well-meaning imaginations. Now that the house is completed, however, now that we are free from the deadlines of television programmes (whose absoluteness would, I might add, make most of the building industry faint clean away!) it is exciting to realise that it is the most basic, least technological, least active elements in the design which play the greatest part towards the energy-conserving aims of the project. It encourages me to re-affirm what I suggested earlier – that the solutions to many of our domestic building problems will come, not from white-hot technology but from common sense.

10. Experience from the 'House for the Future' Project of Granada TV (Part 2)

D. R. Wilson

EXISTING BUILDING AND SITE

Originally, we thought that a conversion based on an existing terrace house or semi-detached type would be most relevant but we abandoned the idea because:

(a) it would have restricted the range of possibilities and techniques which could be demonstrated, particularly those relating to wind power and solar collection,
(b) we could not afford delays over planning permission which might occur in the case of irregular proposals in an otherwise regular situation,
(c) the complications of a combined building and filming operation in a street, at intervals over a long period, were thought to be too difficult

We agreed to consider any suitable existing house or other building of an appropriate size and drew up a list of ideal characteristics for building and site. The one we chose was rather less than ideal, except in terms of location and size in both of which it was very satisfactory, but we were forced into a decision by programme schedules.

The building was a disused coach house, 12 metres by 6 metres in plan and 5 metres high to the eaves, with large openings on the south west side at ground floor level. Other wall openings were generally small and there were none at all on the north east side. It is positioned at the north west end of an approximately rectangular 840 m^2 site which lies near the top of a low hill and slopes approximately 1:15 to the north east. A larger two-storey house, 20 metres away to the south west causes some overshadowing in winter and causes turbulence when the wind blows from that direction. The ground is mainly light sandy loam.

Ground floor walls were 340 mm thick and first floor walls were 225 mm thick, both solid brickwork. The only internal partition was a 225 mm thick brick wall dividing the ground floor into two squares on plan. The roof was covered with Welsh slates on battens, rafters and purlins, and two modified trusses which had been so weakened that they had sagged and had pushed out the top of the south west wall by 90 mm in the middle of its 12 metre length. The first floor was boarded over heavy timber joists (span/depth = 10/1) which run along the length of the building and are supported by the end walls, the centre dividing wall and two timber beams. The ground floor was a mixture of stone paving slabs and clay paviors laid on sand. The condition of the external walls varied from good to very poor depending on aspect and provided useful information on prevailing wind and weather. The roof needed reslating and most window frames and other external joinery needed replacing. There was no damp-proof course.

The time available for research and design was, by any standards, very short though both continued through the whole period of building. We used the three months of programme planning to seek information about similar projects and about the kinds of techniques with which we expected to become involved. We sought advice from universities, research establishments and individuals. For a number of reasons, design work didn't effectively begin until March 1975 and because the television company wanted to start filming in May, we had two months to settle all major issues and design the building in sufficient detail to gain planning permission and building regulation approval. Decision making, at least in the early stages, was largely intuitive. Where we hadn't sufficient experience to make intuitive decisions, we sought it from others and were generally rewarded with co-operation and enthusiasm. We received particular help from Mr A E Mould and Dr J Siviour of the Electricity Council Research Centre, Capenhurst, and from Dr J C McVeigh of Brighton Polytechnic.

We discussed our intentions and established a set of principles on which we could set to work. We agreed to try to evolve a climatically sensible house without preconceptions of its arrangement or its appearance, to attempt to reduce fuel consumption but maintain current comfort standards, and to include, for demonstration purposes, a range of energy saving and energy collecting devices.

We established four areas of consideration:

1) Influence of site and local climate on external form, internal arrangement, and openings for access, daylight and view out.
2) Thermal performance of structure, particularly the external envelope.
3) Ventilation.
4) Power from natural sources.

Climate

Through considering site and local climate we decided that:

1) All habitable rooms including bedrooms should be on SE or SW sides.
2) Areas of greater activity should be located to the north.
3) North facing sides should be lined with storage as additional protection.
4) NW and SW faces of the existing building should be protected from the westerly prevailing wind by lean to extensions for storage and semi outdoor working and living space.
5) Entries should be located to provide access on a lee side in any wind condition.

Thermal performance

Through considering thermal performance of structure we decided that:

1) We were content with a thermally heavy building and that we would try to exploit it rather than change it.
2) The existing solid walls would be insulated externally and clad in a suitable material. This decision was reinforced by the fact that this would prevent further deterioration of eroded brickwork and eliminate wind chill from wet porous walls.
3) The ground and first floors would be temperature zoned and that selected rooms eg: the workshop, would be insulated from adjoining rooms.
4) The existing roof construction would be modified to allow a greater thickness of insulation to be installed.
5) The ground floor and the base of existing walls would be insulated and sloping ground around the building brought up as high as the regulations and simple construction would allow.
6) Openings in the enclosure would be kept as small as the regulations would permit except for the existing large openings in the south west wall.
7) Openings would be fitted with shutters and with external blinds for control of solar gain, and that external horizontal shading would be provided over the large openings.
8) We should be careful to avoid the possibility of overheating in summer and of heat build-up in the heavy, externally insulated, thermal mass.

Ventilation

Through considering ventilation we decided that:

1) The air change should be reduced to an acceptable minimum and that the internal arrangement should take account of air movement from windward to the lee side.
2) To provide a mechanical ventilation system with a heat recovery facility for use when outside temperatures precluded the opening of windows.
3) All openings to be draught-stripped and fitted with a means of fine control over opening movement.

4) Entry into the building should be through a large volume porch and that doors be narrow and arranged to avoid excessive draught.
5) An optional high rate of ventilation should be available, particularly in the conservatory and at first floor level, to cope with the possibility of overheating in summer.
6) Attention should be given to air flow both through any roof collector and across the under-surface, as an alternative to providing external blinds over the collector.

Power

Through considering power from natural sources we decided that:

1) Because water power was unobtainable and methane generation on an individual scale presented difficulties, we should concentrate on sun and wind power only.
2) We would use as much information from existing examples and design, on a largely empirical basis, a solar collector, storage and distribution system, and a windmill, all of which could be made by the clients and other volunteers. The decision to do this followed naturally from the intentions of the programme but, ironically, had to be quickly modified to suit the way in which the television requirements were developing. The eventual arrangement was that I would design the collector, with advice where I needed it, that ECRC would design the storage and distribution system, and that Brian Trueman would design the windmill using a rotor designed by Dr W Cartwright of UMIST and with advice where he needed it.
3) We would experiment with other forms of collectors and storage using air and chemicals.

The arrangement which resulted from these decisions is shown in outline on Figs 1, 2 and 3.

Figure 1 **Ground floor plan.**

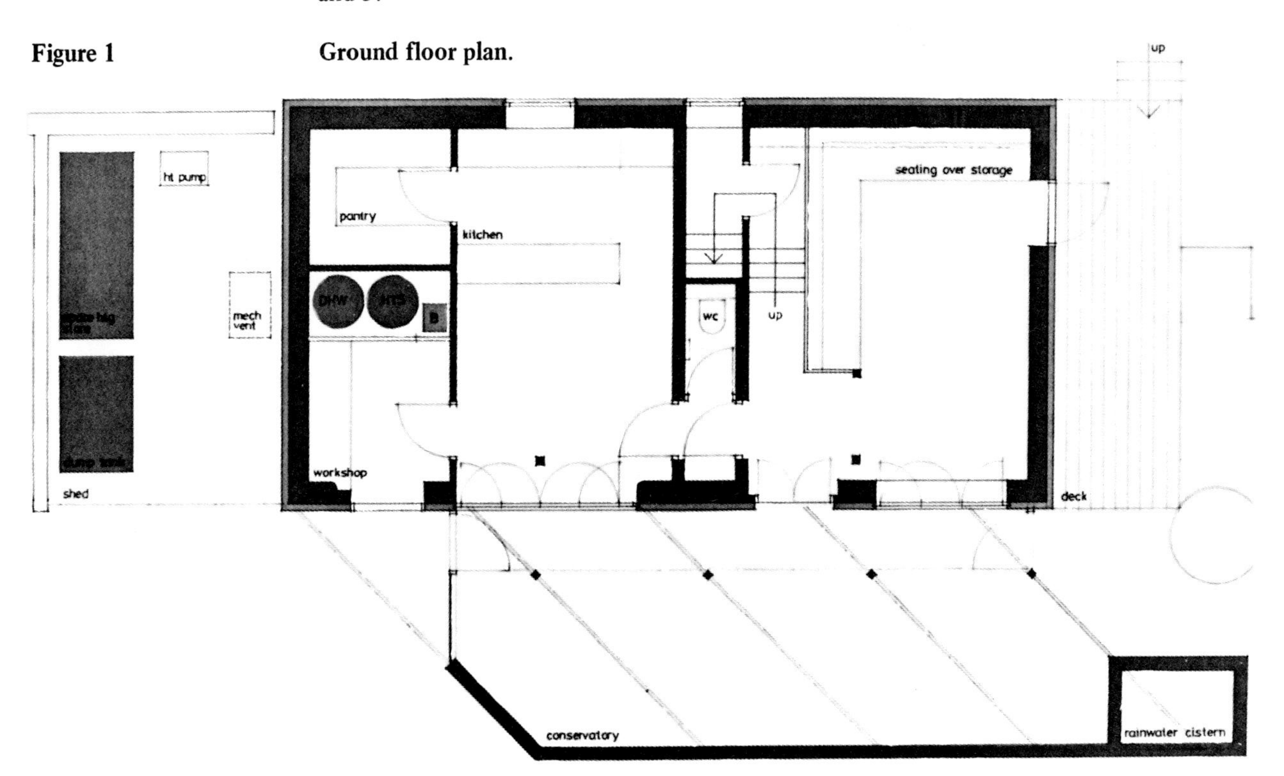

INSULATION AND CLADDING

100 mm thick insulation was provided over the external face of the walls using a variety of materials to test their suitability for fixing by amateurs. We found that, in this respect, semi rigid slabs of mineral wool or glass fibre was the most satisfactory. We were apprehensive about heat build up between the face of the insulation and the back of the cladding, and we decided that lapped cladding such as hung tiles or slates, or weather-boarding would be most satisfactory. We chose sawn tapered section redwood weatherboarding for

Figure 2 **First floor plan.**

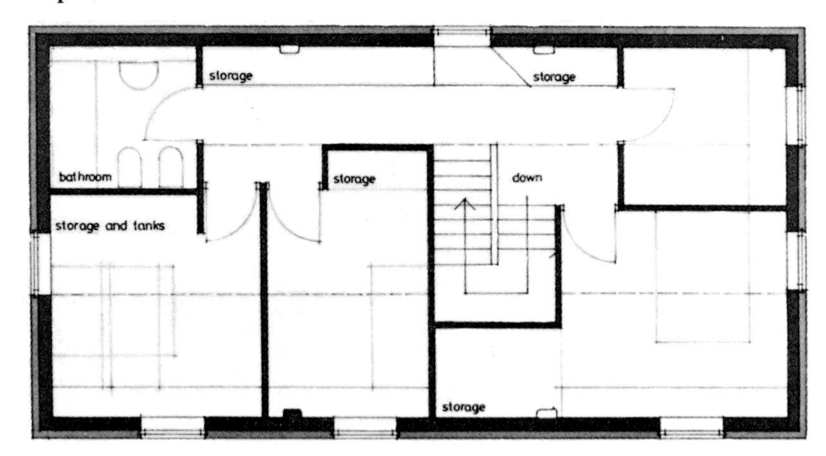

low cost and ease of construction. The material was pressure impregnated with preservative and will be finished with a water-repellant preservative treatment.

Other insulation

50 mm thick insulation was laid between joists in the suspended first floor and the staircase designed to incorporate a self-closing door to prevent air movement in winter from ground to first floor. 100 mm thick insulation was provided over the workshop area and in the partition between workshop and kitchen to prevent excessive heat loss when the external door to the workshop was opened in winter. 50 mm thick extruded expanded polystyrene was installed under a new concrete ground floor slab and turned up to separate external walls from the slab. The same material was applied to the outside of the wall from a level 200 mm above ground down to the top of existing foundations. 150 mm thick insulation was installed in the roof. We adopted the principle of using glass fibre or mineral wool only within the building.

Calculated thermal transmittance in W/m^2K

External walls	0.36
Roof and solar collector	0.22
Ground floor	0.36
First floor	0.5

Figure 3 **Section**

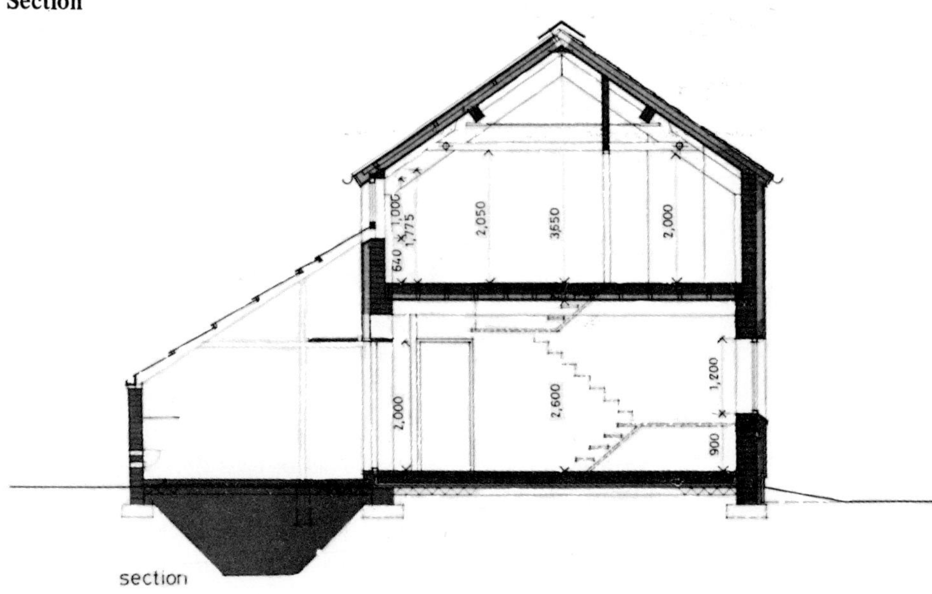

VENTILATION

All openings, except those into the conservatory, are building regulation minimum size and constructed as projecting top-hung (cam-hinged) single sashes. This provides maximum opening area for maximum perimeter length and gives natural ventilation through the top and bottom of the opening. A simple canvas blind can be fed through the top slot to cover the external face of the window in an open position. The window can then be practically closed or remain open to provide ventilation while controlling solar gain. Window frames are specials, made by the general contractor who provided support for the otherwise amateur workforce, and use simple but robust redwood sections with large rebates and neoprene draughtstripping. All window and door frames are fixed in the same plane as the external insulation to reduce cold bridging to a minimum.

A series of high level flaps is provided on the internal face of the first floor ceiling, immediately under the ridge ventilator, to provide an optional natural air extract and to induce airflow across the underside of the ceiling under the solar collector on the SW slope, should it be required. The flaps are insulated and fitted with neoprene draughtstripping for cold weather conditions.

An Aldes mechanical ventilation system was installed with the exchanger, mixers and filters located in the loft space of the shed at the NW end of the house. Two horizontal ducts extend directly from the equipment on either side of the building through the space between the existing longitudinal joists at first floor level. With the exception of two extracts at high level on the first floor, all inlets and extracts are short branches less than 250 mm long. Air is extracted from the kitchen, bathroom and ground floor WC and from the first floor landing. The air is passed through the heat exchanger and heat transferred to incoming air which is fed through the mixers to the sitting room at ground floor and to the bedrooms. Dehumidification of air occurs in the exchanger.

The conservatory serves as an entry with doors at each end allowing access from the lee side. The north west door is the natural entry point for most purposes. House doors are designed as double or multiple leaves to provide the options of narrow opening in winter and wide opening in summer. A continuous ridge ventilator provides weatherproof extraction of air from each bedroom and from the stair and landing area. Insulated internal flaps are provided to close the system off in winter. The ridge ventilator is connected to an air space within the insulation beneath the main solar collector to induce air movement in summer. Flaps are provided at eaves level to close this cavity in winter and from an additional air space. Some of this is shown on Fig 4.

COLLECTOR, STORAGE AND DISTRIBUTION

The collector is a large, low efficiency, 'open-trickle' type extending over the complete 44 m^2 south west slope of the roof. It is 34° pitch and faces 2° south of south west. It was made in this way because:

(a) The existing roof was in poor condition and needed replacing anyway.
(b) It provided an opportunity to show how a collector might be integrated into the total form of the building and not merely 'tacked on'.
(c) The components and operation were all visible and well suited to demonstration on film.

Orientation was recognised as not ideal but sufficiently representative of most existing houses as to be relevant to the programme. We achieved planning permission without difficulty but building regulations application was refused because one clause (clause E17 current regulations) prohibited glass roofs, or glass as a roof cover, over habitable areas less than 6 metres from a boundary. The authority subsequently granted a waiver.

Construction of the collector is shown on Fig 4. The storage and redistribution system is shown on Fig 5.

Figure 4 **Solar roof construction. Drawing by Brian Watson from "A House for the Future" by T P McClaughlin by permission of ITV Books Ltd London.**

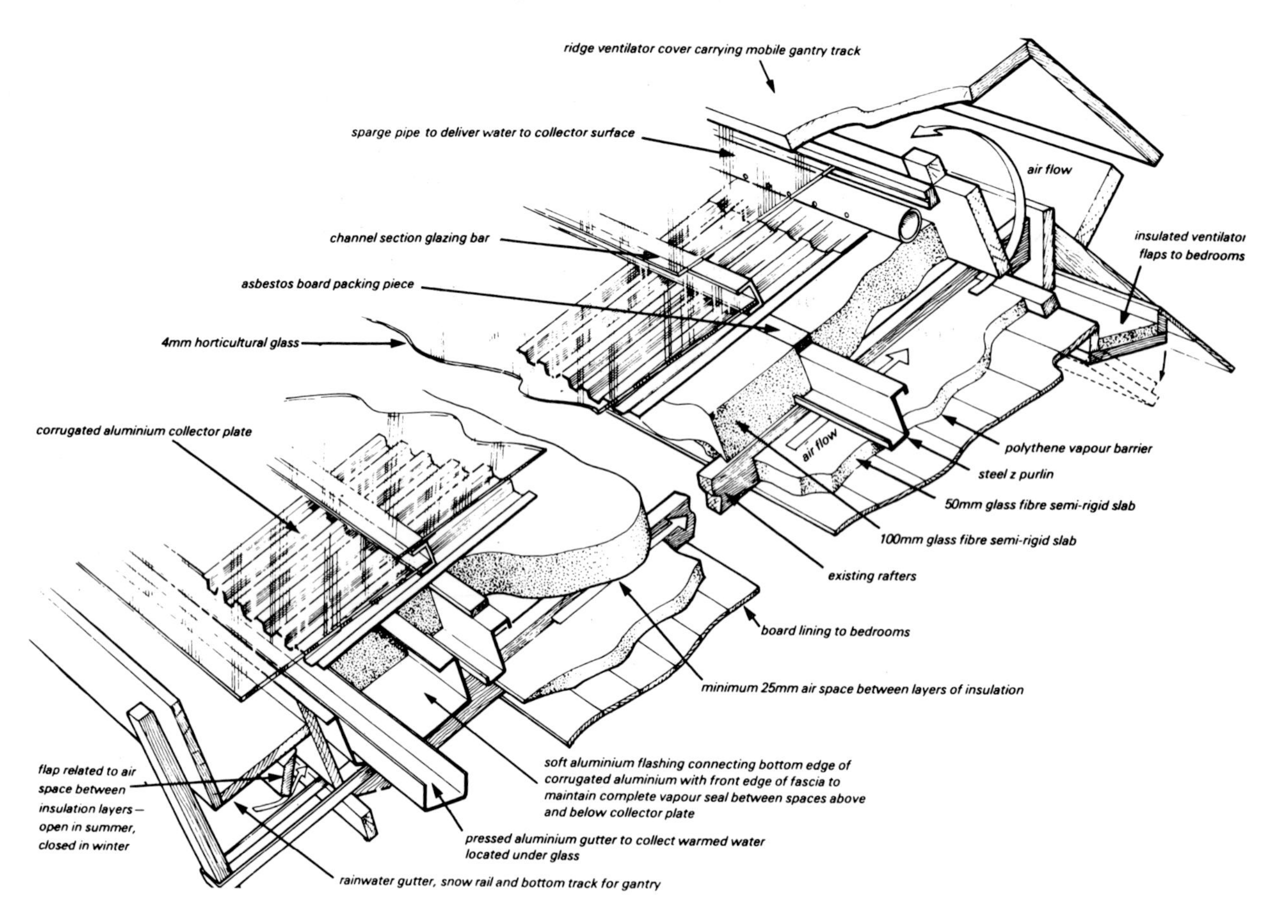

Heat from the roof collector is transferred into storage through 2000 litres of water and inhibitors circulating through a pump to a sparge pipe at the ridge, over the corrugated collector plate, into a gutter below the glass cover at the eaves, through two heat exchangers, and back into a reservoir called the dump tank at ground level. Automatic control, by means of sensors and motorised valves, diverts water to the 400 litre domestic hot water cylinder at 60°-45°C and to the 3000 litre space heating store at 45°-25°C. From 25°-5°C, water circulates only through the dump tank and heat stored there for transfer at 25°+ into the space heating store by means of a heat pump using off-peak supply.

Additional heat is supplied to the domestic hot water cylinder through two 24 volt/600 watt immersion heaters to be powered by the windmill, and by a standard domestic 240 volt immersion heater for conditions of no sun and no wind, and as a means of 'topping up'.

Additional heat can be supplied to the space heating system by mixing in water at maximum 95°C from a 700 litre storage tank heated by a 3.5 kW solid fuel boiler. Mixing is thermostatically controlled through a motorised mixing valve.

The 2000 litre dump tank and the 300 litre space heating store are located outside the house to prevent overheating the house interior in summer. Tanks are enclosed in a minimum of 150 mm thick insulation.

THE CONSERVATORY AS A COLLECTOR

The glazed roof of the conservatory is arranged to meet the main wall of the house 150 mm above the cill of the first floor windows, so that in favourable conditions air warmed in the conservatory can be admitted to the bedrooms through a 100 mm high flap ventilator running the width of each window. Inlet flaps to the conservatory are provided

Figure 5 **Simplified diagram of solar heating system.**

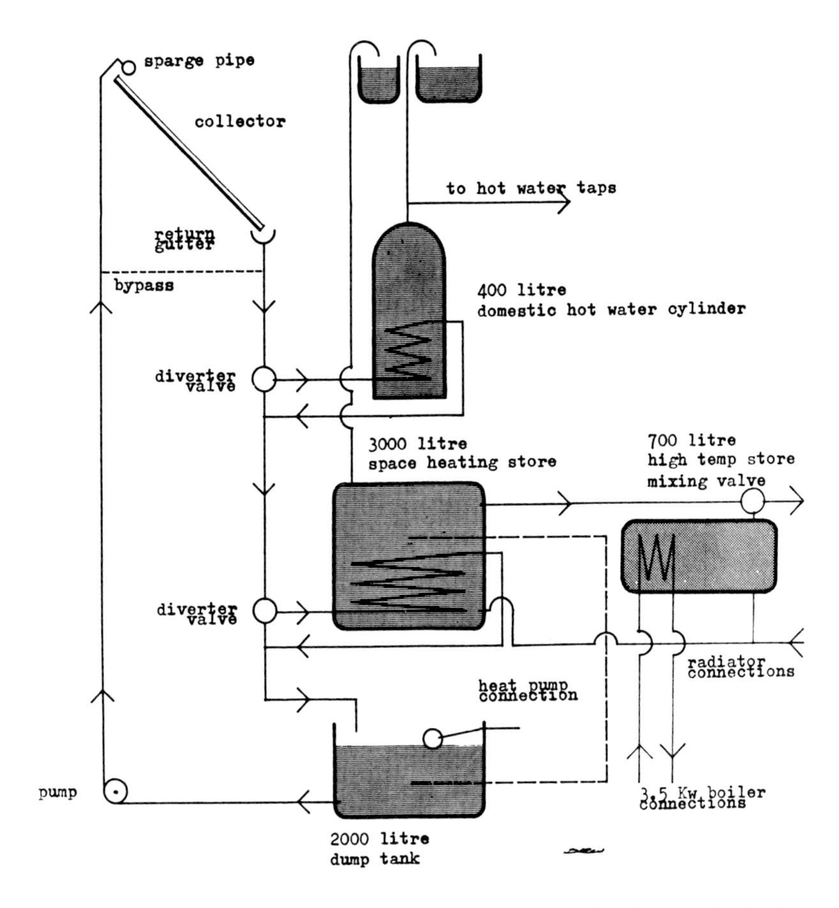

at low level under the level of the greenhouse staging and outlet flaps to the bedrooms are provided for reasons already described.

Storage of excess heat is being attempted by installing 12 m^3 of crushed limestone, size graded 75 mm, in a pit below the conservatory floor level. The crushed stone is insulated from the ground by surrounding it with 50 mm thickness extruded expanded polystyrene but not from the stone paving over. Fig 6 shows the sectional form of the store and the general arrangement. There are three possible ways of redistributing heat:

(a) By conduction through the paving and slow radiation from that surface with the fan off.

(b) By removing the lid of the chamber to encourage slow reverse air flow to occur with the fan off.

(c) By removing the lid of the chamber and running the fan at slow speed in reverse.

SUMMARY

The planning arrangement appears to be generally satisfactory with the exception of the workshop and shed which have been invaded by a substantial volume of storage tanks and plant. The usable space remaining is disappointingly small.

External insulation and cladding was relatively cheap and easy to apply. Main problems were fixing the timber battens to soft brickwork and maintaining accurate batten centres for the polystyrene insulation. The development of an improved supporting technique for cladding over external insulation is badly needed.

The weatherboarding was fixed completely by amateurs and is generally satisfactory except for the SW elevation which had a tendency to warp during the hot summer of 1976 but recovered during the following winter. Windows operate satisfactorily and the external canvas blind is effective. We couldn't find a window catch with a satisfactorily fine control

Figure 6 **Conservatory heating system .**

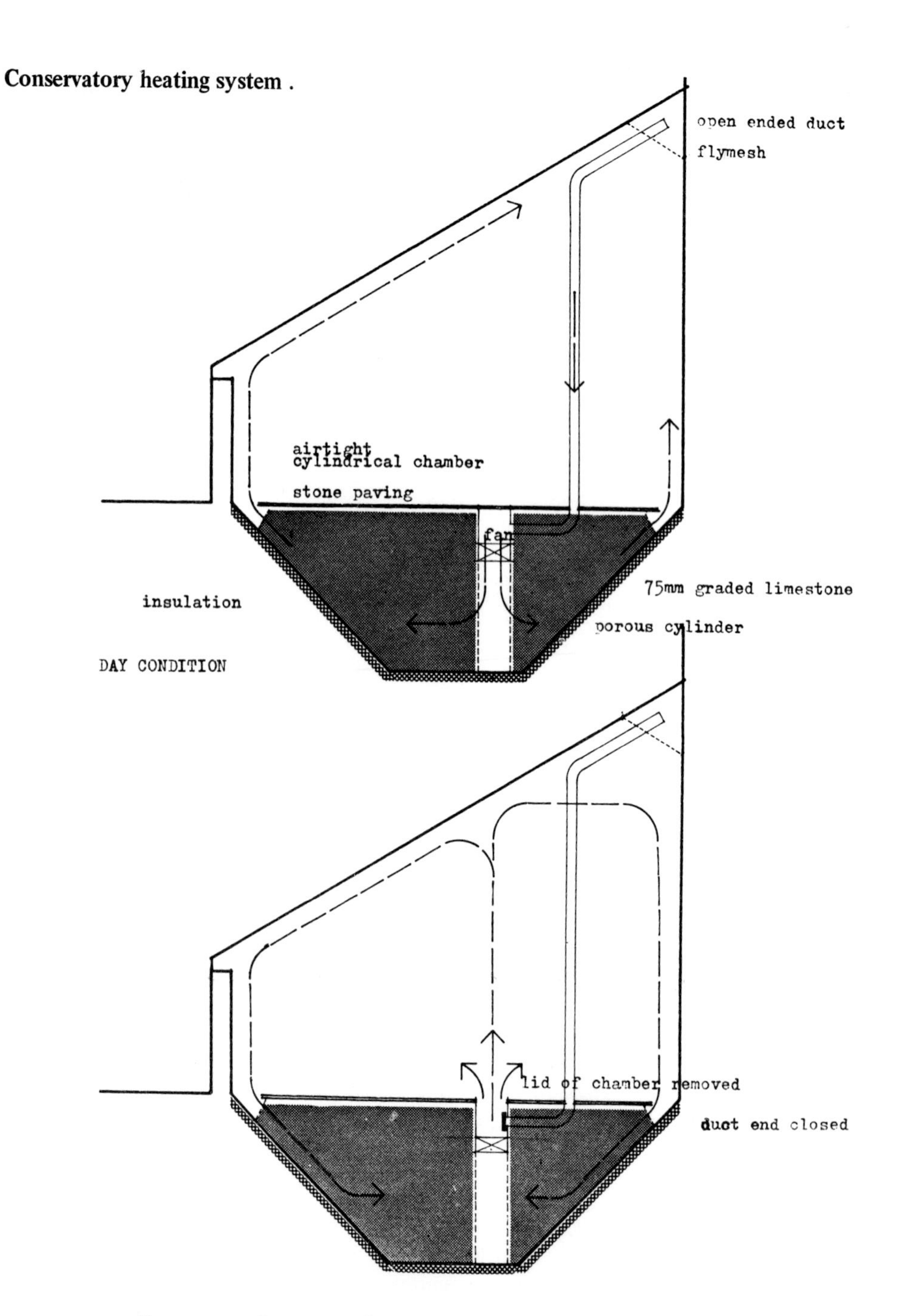

over very small extents of opening (screw adjustment for example) and had to make do with a standard cam fastener.

The high level flaps under the ridge were kept open day and night during the period May – September for maximum comfort and have been closed all winter. Intermittent operation appears to be unnecessary.

The integral roof collector was extremely difficult to construct, almost entirely due to problems caused by trying to fit a precise, regular structure to a very imprecise, irregular structure. In retrospect, the fit we tried to achieve between the two was too tight. There is a considerable amount of information and experience available as a result of our work on this component. The structure at present performs satisfactorily. There has been no discomfort through overheating during the summer of 1976 and no discernible radiation from the underside of the collector over the bedroom area even when the collector has been left dry for long periods of up to two weeks. Provision for movement appears to be satisfactory and there is no perceptible noise from the collector either from water or from structural movement.

Thermocouples were installed by the Electricity Council Research Centre during the construction work and are located at the interface of separate materials in the roof, walls and ground floor slab. The performance of the house will be monitored over a two year period. The house has been occupied since May 1976 by Mr and Mrs Grant who acted as clients for the project and, with Brian Trueman, collaborated in its design.

DISCUSSION

Dr. D. Fitzgerald (University of Leeds)

Mr Wilson said he has used closed cell expanded plastic foam insulation underground and that it is waterproof. It is indeed proof to liquid water but experience in district heating has shown that these materials are not proof to water vapour. Water vapour will diffuse from one cell to the next – depending on the water vapour pressure differentials and condensation will take place, depending on the relation between the concentration of the water vapour and the temperature, the thermal resistance will then disappear and the material will become a sodden mass. Anyone in this field would be well advised to heed the strong warning against the use of expanded foam insulation underground, unless a vapour proof membrane is provided. Without a vapour proof membrane there will be trouble – not in the first year or the year after – not maybe even in five years – but in time the insulation will fail. This applies to anything underground, because the water table can at times be anywhere from above ground level downwards.

D. R. Wilson

The polythene damp proof membrane on top of the insulation would help because it would act as a vapour barrier.

J. Owen Lewis (University College, Dublin)

I wonder if it is possible that there is confusion between two distinct polystyrene materials, one an extruded polystyrene and the second the more common expanded polystyrene. Extruded polystyrene (which has a closed cell structure) has been used for quite a number of years in Germany and is increasingly used in this country in "upside-down" roof constructions where it is repeatedly exposed to moisture.

D. R. Wilson

No expanded polystyrene was used within the house at all except under the ground floor slab, because just about all fire officers are against it. We did not want to provide it in any situation where electrical conduits or fittings or cables might come into contact with it.

11. Experimental Low Energy Houses at the Building Research Establishment

K. J. Seymour-Walker

INTRODUCTION

The major study of energy usage in housing in the UK (1) undertaken by BRE in 1974/5 identified the measures which could contribute to savings in the country's primary energy consumption. Some experimental houses are now being constructed, within the BRS site at Garston, which will demonstrate the application of these measures. They will be fully instrumented and their behaviour will be observed under various controlled static conditions, and under simulated occupancy.

At the outset it was decided that these experimental houses should:

not require unfamiliar construction techniques;
use only realistic technology;
not require any major change in the inhabitants' life-styles;
be suitable for normal urban/suburban sites.

There would be no attempt to make them independent of the normal public utility services. As a result of these decisions, the houses do not present any revolutionary image; however, the energy savings they represent are in fact highly significant.

Of the numerous energy-saving possibilities, some are mutually exclusive, and it was felt that more than one experimental system was called for. Initially three houses have been planned and, for convenience, they are known as the Heat Reclaim, Solar Energy and Heat Pump houses.

HOUSE SHELLS

The decision to use familiar construction techniques led to the choice of a timber-framed building system for some of the houses. The techniques of this type of building are now well understood, and the construction readily permits an increase in thermal insulation. It happened that an estate of timber-framed houses being built at Bretton, Peterborough, was already well-known to some of the team, as studies of district heating were being conducted there. Thus the plan-form of the Type 47 house from Bretton was adopted for the Heat Reclaim and Solar Energy houses (Fig 1).

These are conventional 5-person terrace houses, 5.7 m frontage, 8.6 m depth, with dining-kitchen, living room, cloaks and 1.7 m^2 store downstairs, 3 bedrooms and bathroom upstairs. Three such houses are being built as a terrace, on a site within the grounds of the Building Research Station. The two end ones (effectively semi-detached) will be the experimental houses while the central one forms a thermal 'buffer' and houses all the monitoring instrumentation.

The timber structure is prefabricated into large panels as the 'Frameform' system. Exterior cladding is facing brick on ground floor and flank walls, weatherboarding on first floor front and back. The only change from the standard 'Frameform' is that the thermal insulation in the panels and roof is increased from 25 mm to 92 and 100 mm of glass-fibre respectively. This gives a U-value of about 0.29. The roofs, covered with felt

Figure 1 **Plans of terrace of experimental houses.**

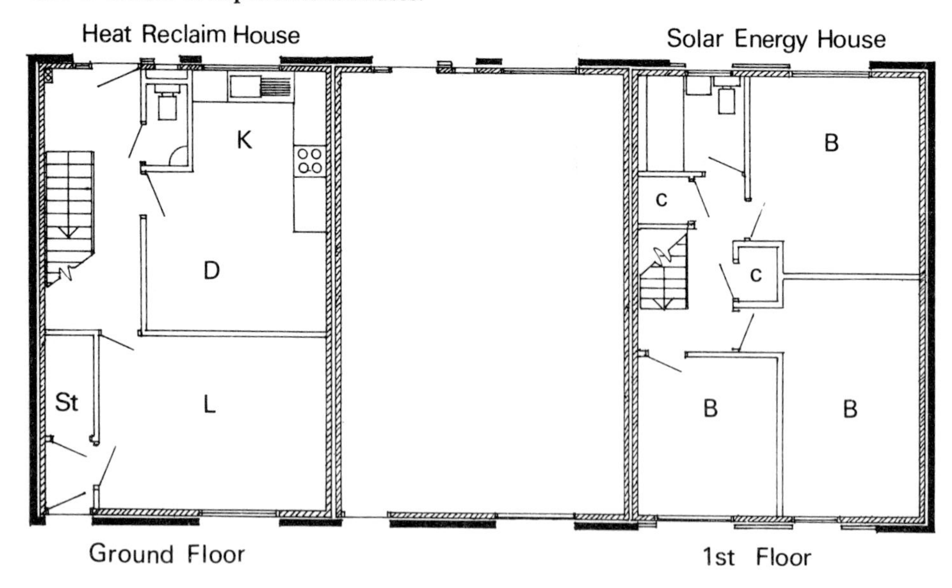

Figure 2 **Cross-section of heat pump house.**

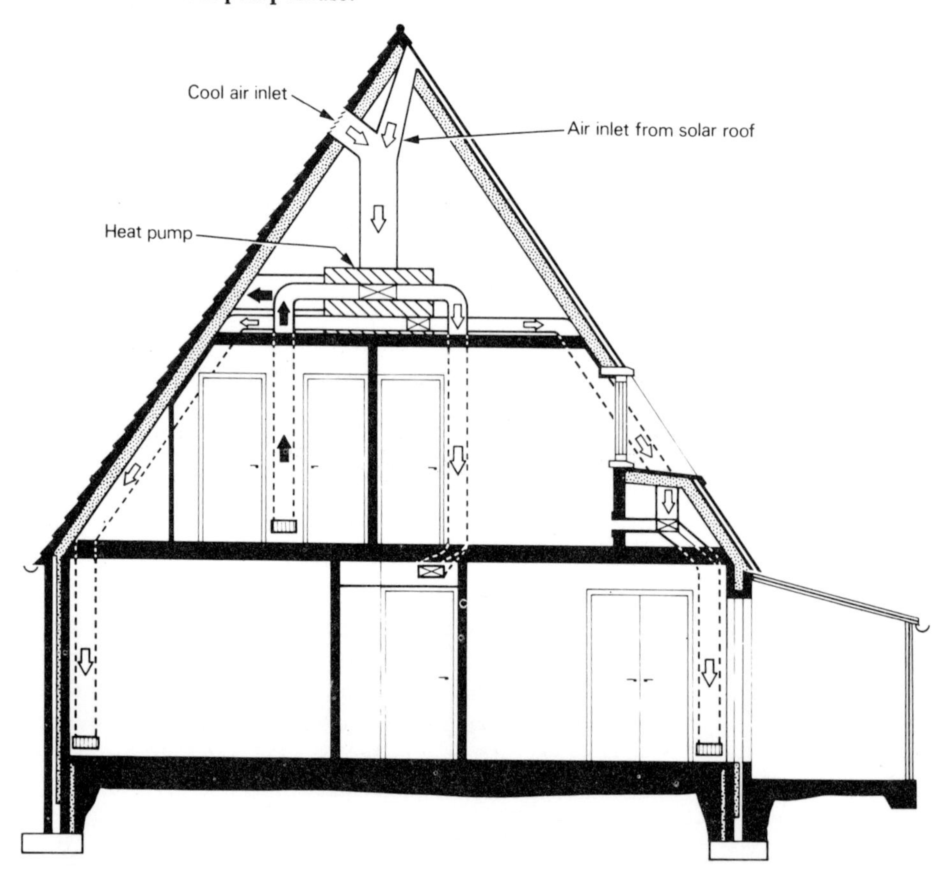

tiles, are of 42° pitch; this steepening, as compared with the 22½° at Bretton, provides a more suitable inclination for the solar collector and increases the loft-space available for experimental access. The windows are initially of the same type as at Bretton, with modifications such that the openable portions will have very low infiltration rates when properly closed. Windows of higher quality, or double glazing, may be fitted later.

The architectural design of the Heat Pump house, which is built alongside the others, is of 'chalet' type with upstairs rooms accommodated within the pitched roof, and intended to be built in semi-detached form (although a terrace variant is possible). The envelope area is small relative to the habitable volume (Fig 2). This house is of loadbearing cavity

Figure 3 Ventilation and hot water recovery systems – Heat Reclaim House.

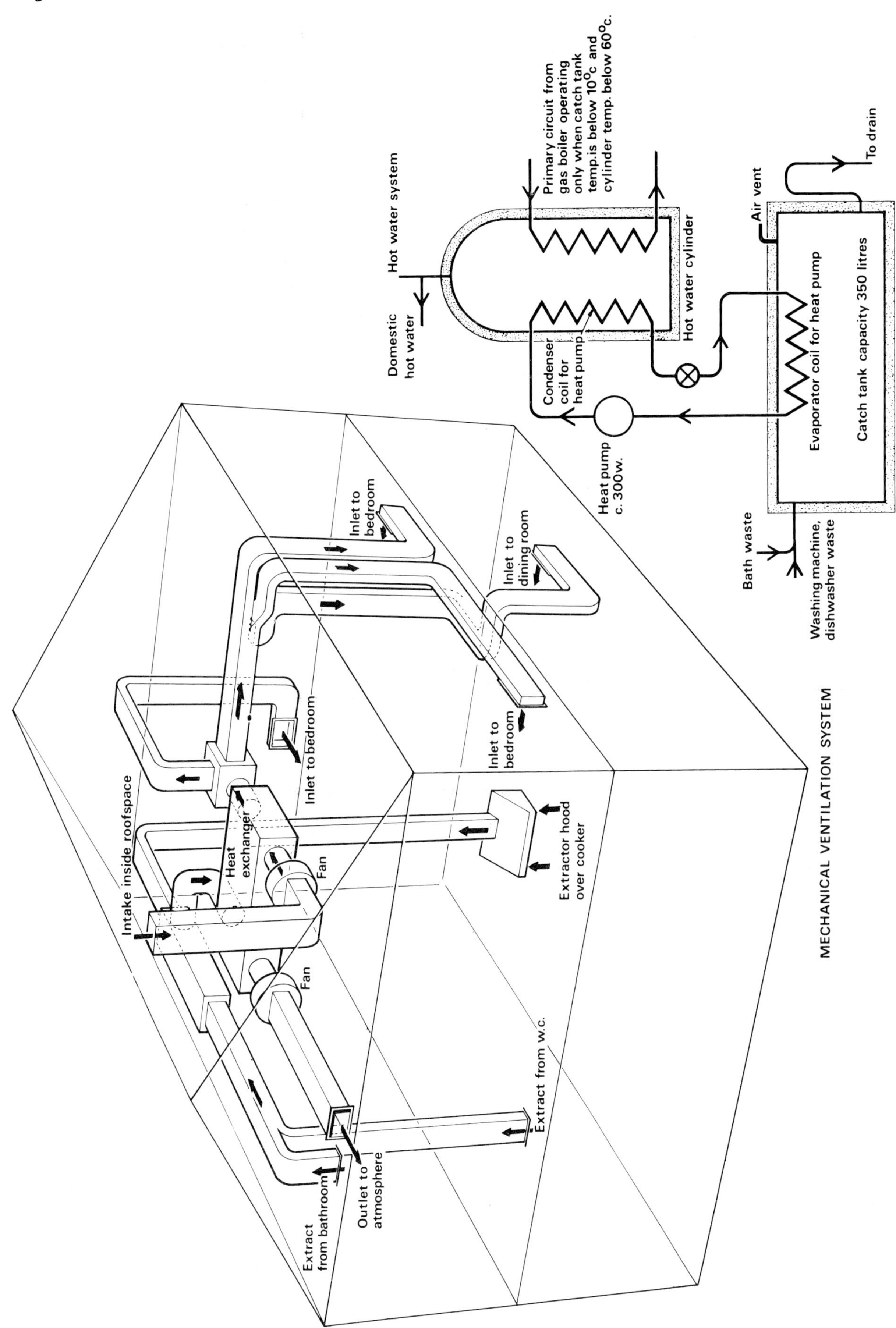

wall construction; the outer leaf is facing brick, with cavity, rigid plastics foam, lightweight concrete block, and foil-backed plasterboard on battens as the inner dry lining. This wall construction also has a U-value of 0.29. Openable windows are used, but the area of glazing has been kept small, except for the ground-floor on the south side where French doors open into a glazed conservatory.

The ground floor is an insulated solid slab, the first floor of timber and the internal partitions of conventional block and plasterboard construction. The roof has a steep pitch (54½°) for aesthetic reasons and to suit the chalet layout. It also provides a space for the heat pump machinery, and its south-facing slope forms a simple solar collector, being corrugated, blackened, aluminium sheet. Air passes underneath this sheeting on its way to the evaporator coils of the heat pump.

The internal planning of the house, with a wide central hallway and one of the bedrooms downstairs, reflects the proposals for adaptability put forward by Rabeneck, Sheppard and Town (2). The floor area (99 m^2) is somewhat larger than that of the timber houses (88 m^2).

MECHANICAL SERVICES

Heat reclaim house

Heat reclamation is applied both to space heating and to the hot water supply. Although openable windows are provided, there is a mechanical system capable of handling the

Figure 4 **Systems diagram of the experimental solar house.**

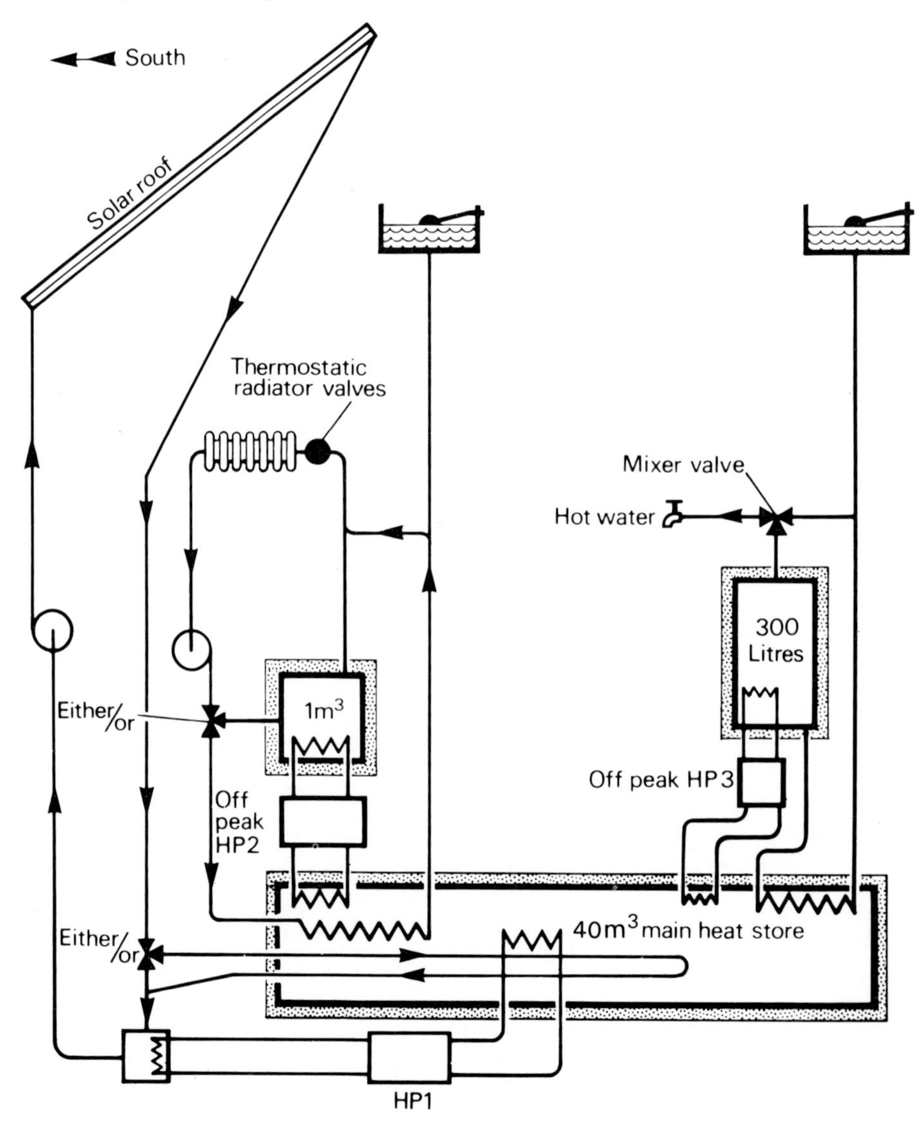

whole of the ventilation requirements of the house during the heating season. One fan in the roof-space extracts air from toilet, bathroom and kitchen – the latter via a hood over the cooker. Another fan draws in fresh air and feeds it through ducts to all main habitable rooms. The in and out streams pass through two sides of a heat exchanger, so that much of the heat in the outgoing air is transferred to the ingoing (Fig 3). Without this heat recovery the ventilation heat loss would be about 44% of the space heating load. A heat-exchange efficiency of about 60% is expected, using a proprietary set of components of French manufacture.

The heating system itself will have normal hot-water radiators, supplied by a low-capacity lightweight gas-fired boiler. The fresh-air outlets of the ventilation system are in the walls behind the radiators, so that the incoming air is immediately warmed to avoid cold draughts.

The hot water system also uses heat reclaim. Waste water from bath, dishwasher and washing-machine outlets is led into a storage tank (1 m^3) before discharge to drain. Sited near the top of this tank is the evaporator coil of a small heat pump (200 W) whose condenser coil is in a hot-water cylinder. The heat recovered by the heat pump, with the energy used to drive it, is expected to provide about 70% of the hot water requirement. The remainder can be supplied by a normal coil in a second cylinder, on the boiler primary circuit. The control system will ensure that boiler heat is only called in when the heat pump's output is inadequate.

No attempt is made to reclaim any heat in waste water from washbasins or from the kitchen sink; the amount of heat which could be reclaimed from these is small, and there might be difficulties with grease and scum collecting in the storage tank.

Solar energy house

This house, of similar plan to the Heat Reclaim house, does not have mechanical ventilation. Its energy system (Fig 4) is based on a solar collector, of established design and 20 m^2 in area. The south-facing pitch of the roof is covered externally with patent glazing. The solar collector elements are installed below this, so that they can be serviced or replaced without disturbing the weather-tightness of the roof.

During the summer the collector should deliver more than enough energy for space and water heating. Excess energy can be stored by heating water in a 40 m^3 well-insulated tank and any useful solar energy available in the winter can also be passed to this.

The space heating is provided by hot water radiators fed from microbore piping. The radiators are of the extended-surface type so that lower water temperatures can be used. They are supplied with heat from the 40 m^3 tank via a heat exchanger when its temperature is high enough. At other times the supply is drawn from a 2.2 m^3 well-insulated tank which is heated during off-peak hours by a small water/water heat pump using the 40 m^3 tank as its heat source. The heat exchange loop would still be used on start-up to preheat the radiators from cold.

Water flowing into the hot water supply system will pass through a high capacity heat exchanger coil in the 40 m^3 tank to a 300 1 storage tank, sufficient for a day's consumption. A mixer valve will ensure that the outlet temperature does not exceed 60°C when the 40 m^3 tank is hot. When it is cool, another off-peak heat pump will use it as a source to heat the 300 1 tank.

The solar input from the collector will be fed into the 40 m^3 storage tank through a heat exchanger when the input temperature is higher than the tank temperature, and via a third heat-pump when it is lower. This last pump can in fact be used as an 'air-source' machine (via the solar-collector surfaces) even when there is no insulation. The control system will limit this mode to situations when the 40 m^3 tank is seriously depleted.

Heat pump house

The equipment for this house comprises three electrically-driven heat pumps, providing space heating, ventilation heat recovery and hot water (Fig 5). The main space-heating

Figure 5 **Systems of heat pump house – winter operation.**

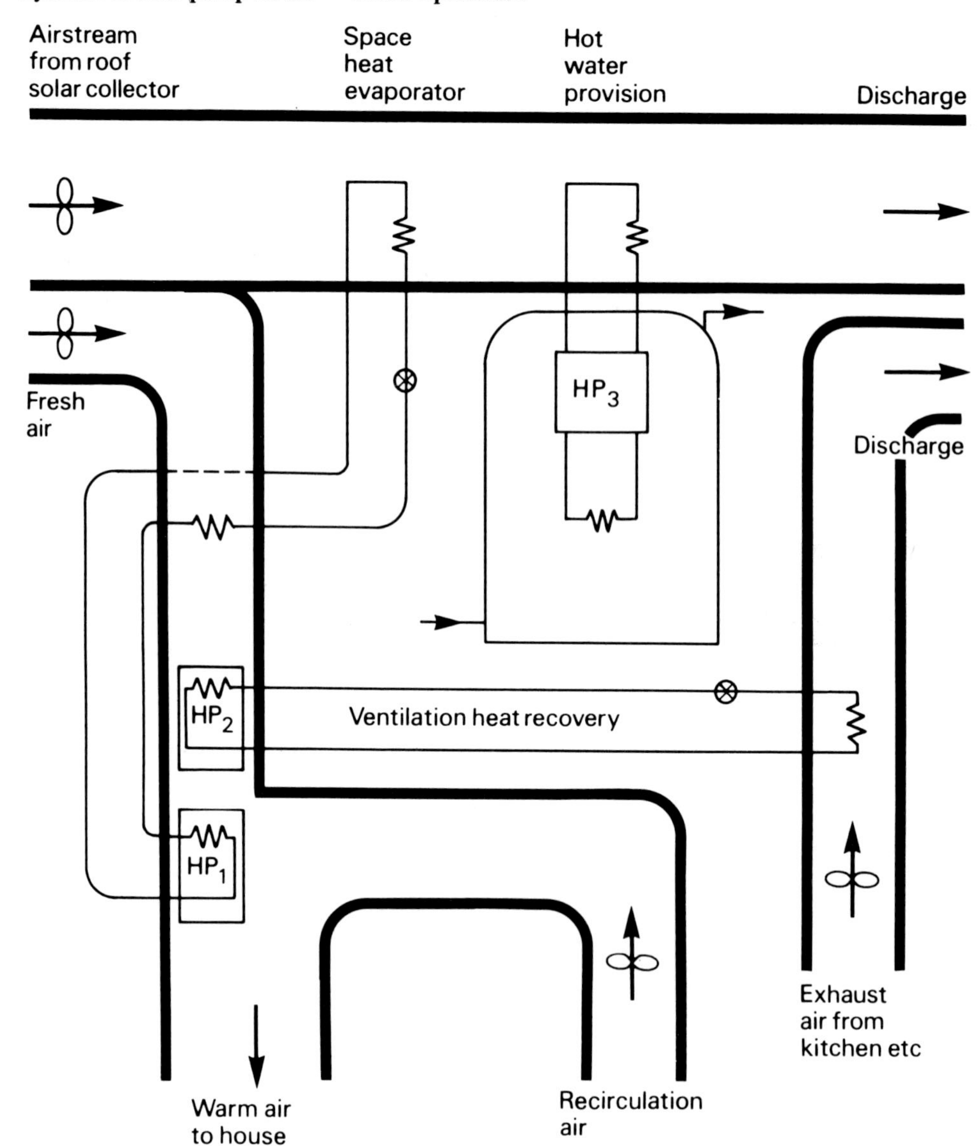

pump is an air-to-air machine, using as its heat source external air which has passed under the 'solar roof'. This air, passing through the evaporator of the heat pump, may be at a temperature appreciably above that of the ambient and certainly not below it.

The mechanical ventilation system has a duty similar to that in the Heat Reclaim House, extracting in the same manner. The outgoing air passes through the evaporator of the heat recovery heat pump, before being ejected to atmosphere. The incoming make-up air has been drawn under the 'solar roof', and then passed through the condenser of the heat recovery heat pump, before being mixed with the recirculating air going to the main space-heating condenser.

Water is heated by a third, air-to-water, heat pump, whose condenser is inside the hot-water cylinder. This pump has two evaporators: one is in the outgoing extract air duct, downstream of the heat recovery evaporator. The other is in the downstream part of the duct taking external ('solar roof') air through the main space-heating evaporator. The pump will draw energy from whichever of the two evaporators is the more advantageous, according to weather conditions (Fig 6).

EXPECTED ENERGY CONSUMPTION

Table 1 shows the estimated energy consumptions in GJ per annum of the space

Figure 6 **Systems of heat pump house – summer operation.**

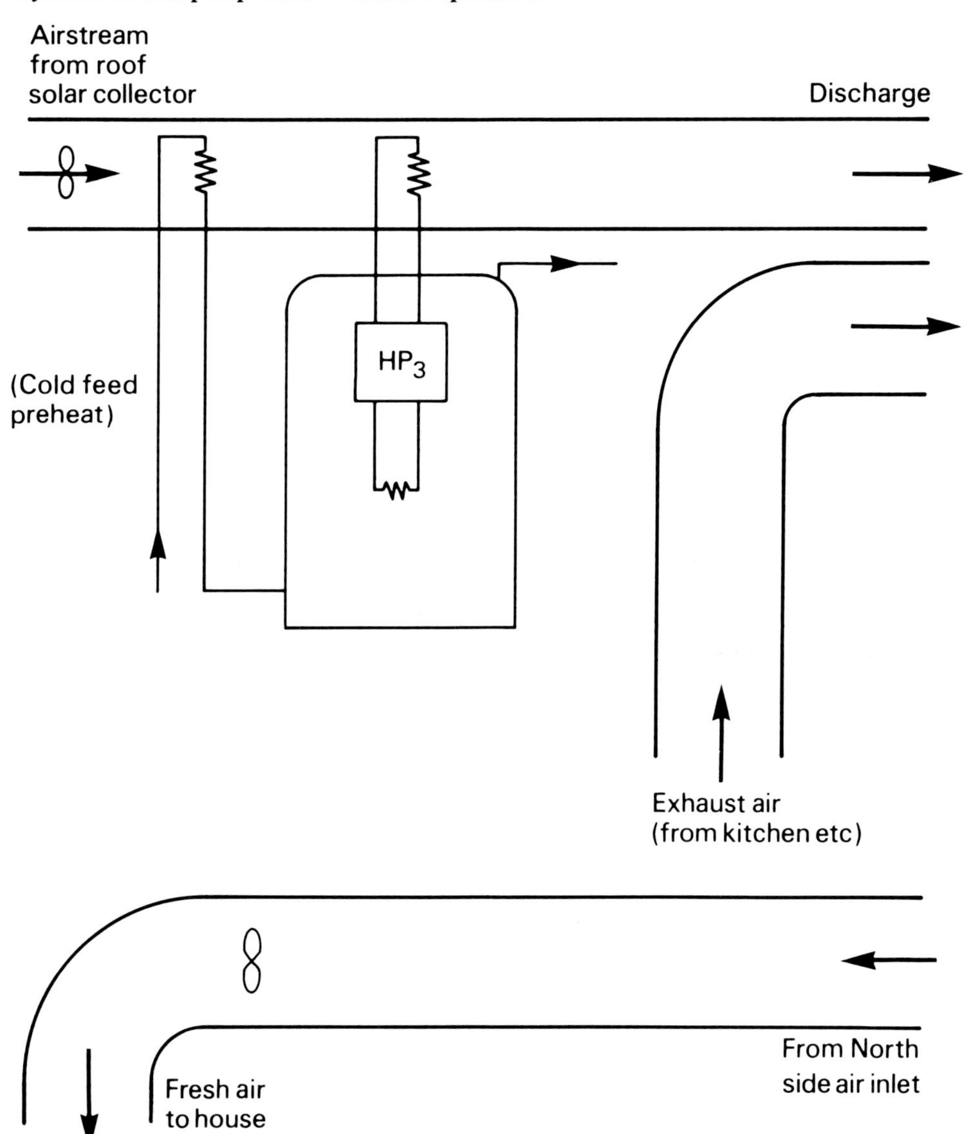

Figure 7 **Experimental low energy houses under construction.**

and water-heating systems, for the experimental houses and two comparison houses. The overheads associated with the supply of the various energies are taken to be such that, of the initial primary energy the percentages reaching the consumer are 27, 94 and 90 for electricity, gas and LPG respectively. Utilisation efficiencies within the houses are taken as 100%, 60% and 60%.

Table 1 **Estimated energy consumptions of the space and water-heating systems.**

		Net Energy	*Form of Supply**	*Primary Energy*
Bretton Type 47 house	Space	54.0	Average mixture*	124.2
to current Building Regs	Water	12.0	" "	27.6
				151.8
Bretton Type 47 house	Space	27.0	" "	62.1
with 0.29 U-value	Water	12.0	" "	27.6
				89.7
Heat Reclaim House	Space	21.0	Gas	37.1
	Water	3.0	Electricity	11.1
	Water	3.5	Gas	6.2
				54.4
Solar Energy House	Space) Water)	13.5	Electricity	50.0
Heat Pump House	Space	8.3	Electricity	30.7
	Space	0.7	LPG (Boosting)	1.3
	Water	5.0	Electricity	18.5
				50.5

*Average mixture: Gas, electricity and oil in the proportions in which they contribute to the national total domestic consumption; average overhead 130%.

All houses will need a further 3 GJ electrical (net) for lighting, TV and miscellaneous electrical consumption, ie 11.1 GJ primary. Cooking will require 10 GJ net if gas (10.6 primary), or 5 GJ if electric (18.5 primary). The heat gains into the house from these have been allowed for in the space heating calculations.

The predictions for the Heat Pump and Solar Energy Houses are cautious (particularly for the latter) pending the completion of more sophisticated simulation studies. In the case of the Solar Energy House various operational strategies are possible and these are still being evaulated, the sizes of the solar collector and the main storage tank having been determined by practical considerations rather than by optimisation studies.

ACKNOWLEDGEMENT

The work described has been carried out as part of the research programme of the Building Research Establishment of the Department of the Environment and this paper is published by permission of the Director.

REFERENCES

1. Building Research Establishment. *Energy Conservation: A study of energy consumption in buildings and possible means of saving energy in housing.* BRE Current Paper CP56/75. Garston: 1975.
2. Rabeneck, Sheppard and Town. *Housing flexibility/adaptability?* Architectural Design 1974, **49**, 2, pp 76-90.

DISCUSSION

Dr. G. S. Saluja (Robert Gordon's Institute of Technology)

Could Mr Seymour-Walker give some idea of the cost of installing a heat pump and other systems in the three houses? How do these costs compare with the conventional central heating system?

K. J. Seymour-Walker

The actual costs of the houses bear no relation to normal production cost. They are one-off prototypes and carry the kind of on-costs that one expects in such circumstances. For the heat reclaim house, the additional hardware, if it was in reasonable quantity production might cost about £600. For the heat pump house the additional cost over a conventional heating system would be of the same order. The additional cost of the solar house is very much higher, probably a minimum of £3000, bearing in mind both the 20 m^2 of collector and the 40 m^3 storage tank. Some savings would no doubt be possible, but as the house stands at the moment it is very expensive.

K. W. Reece (Ormrod & Partners)

The energy problem has been brought to a head by the shortage of oil, amongst other things. Yet many of the papers and much of the discussion has been concerned with the use of plastics for insulation, materials that originate from oil. Should we not therefore be looking for insulation materials other than plastics?

K. J. Seymour-Walker

Because of the extremely long life of a building, the energy used in its construction and in the manufacture of the components, including insulation materials, from which it is made is a very small percentage of the energy use that the building generates. Even in a grossly extravagant construction this does not exceed about 15% of the energy that the building can be expected to use during its lifetime. Some insulating materials are less energy intensive or less oil intensive than others, a factor which may be expected to be reflected in their price, if not now at least in the future.

12. Long Term Heat Storage

J. E. Randell

INTRODUCTION

Since the requirements of space heating and the availability of solar heat are generally out of phase, heat storage is a key aspect of solar heating. It can play an important role in the use of other ambient energy sources. It can be used with advantage in heat pump systems and, because it provides a relatively cheap and simple method of storing energy, might be preferred to a power storage system if wind is used as the source of energy for heating.

With solar heating, storage is generally advocated to enable the supply of heat to be carried over from day to night. It is argued that in this country though, the quantity of heat stored should be greater than is required for this, to allow for periods of low solar radiation availability. It is logical to seek to extend the period over which heat can be supplied from storage to enable heat to be collected in warm periods of the year for use during cold periods. A limit is reached when inter-seasonal storage permits a deficiency of heat in the winter months to be met entirely from heat collected during the summer and stored and the possibilities of achieving this will be considered.

Until recently, the storage of heat at temperatures close to ambient has not been the subject of intensive research. An insulated tank of water presents a simple solution to storage problems and is generally satisfactory. The use of storage tends in any case to be minimised and the reasons for this are apparent. Space is at a premium and capital cost is a prime consideration in the design of any heating installation. Storage equipment is bulky and adds to the cost. The object of this paper is, taking a broad but practical view, to examine the feasibility of long term heat storage and whether it is necessarily too bulky or too expensive.

The storage capacity required is considered first, assuming inter-seasonal storage since this sets a limit to what is desirable. The temperature level at which heat is stored is important, particularly when a heat pump is employed to upgrade heat. In the light of this, possibilities of heat storage using salt hydrates, which are relatively cheap and potentially capable of storing heat with a high energy density, are reviewed. Because it has a large influence on the size and cost of heat storage equipment, the design of heat stores is also considered, from a practical viewpoint. Finally, the storage requirements for specific applications are assessed. The applications are hypothetical but selected as typical of situations in which long term heat storage might be employed and the space requirements and costs arrived at lead to some tentative conclusions.

HEAT STORAGE REQUIREMENTS

If heat storage is to play its full part in solar heating it will obviate the need for supplementary heating and permit solar energy to meet the whole space heating load.

A typical space heating load, for a well-insulated house in the United Kingdom, over a twelve-month period commencing at the end of March is shown in Fig 1 by the histogram and cumulatively by curve 'A'. The quantity of heat required is shown, allowance having

Figure 1 **Heating load and heat collection curves using average values.**

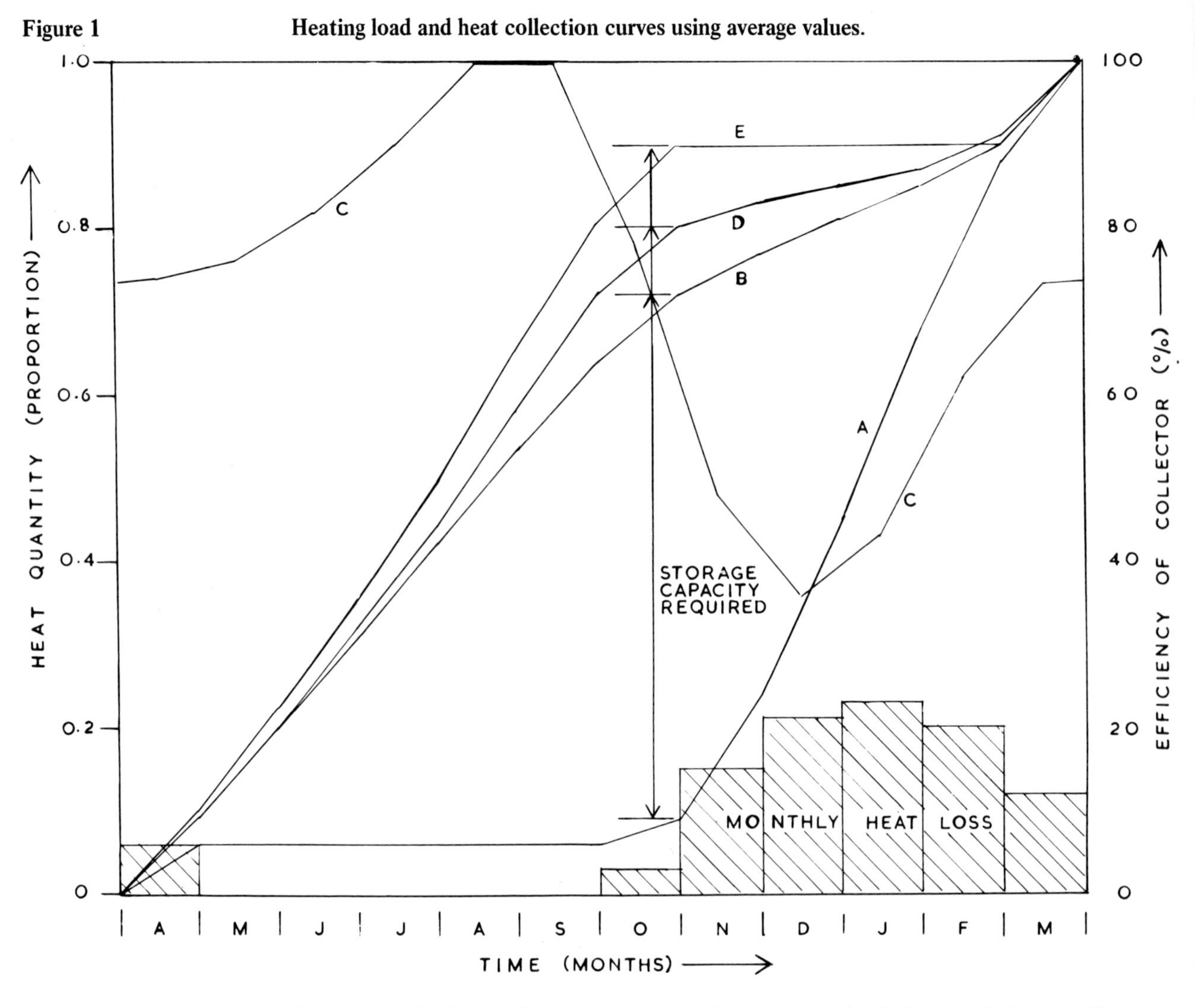

been made for heat gains due to solar radiation through windows and for miscellaneous gains, and the load is expressed as a proportion of the annual space heating load. The total radiation incident on a vertical surface facing South is shown, cumulatively from the end of March and based on average values, by curve 'B'. The rate at which heat is picked up by a solar collector at a particular time will depend upon the type of collector, the outside temperature and the intensity of radiation, besides the collection temperature.

It will be assumed firstly though that the collector has an efficiency of 100% throughout the year and that curve 'B' represents the actual quantity of heat collected as the year progresses. The area of collector required is the minimum, being that for which the total quantity of heat collected over the year is equal to the annual requirement for heating. During the month of March, the quantity of heat collected is less than is required for heating and heat must be drawn from a store whilst in April more heat is collected than is required for heating and heat will be accumulated in the store. It has been assumed, therefore, that the store will be exhausted at the beginning of April and the cumulative heat quantities have been taken from this time for this reason. The maximum quantity of heat that must be stored is represented by the maximum difference between the values indicated by curves 'A' and 'B'. A quantity of heat equal to 64% of the annual heat requirement needs to be stored.

Allowance can be made for reduction in the efficiency of the solar collector in winter but details of the installation need to be known to be precise. For the purpose of comparison, it will be assumed here that the efficiency of the collector varies as indicated by curve 'C' in Fig 1, a value of 100% being taken still at the hottest time of the year. The cumulative quantity of heat collected is then represented by curve 'D' in Fig 1. Over the

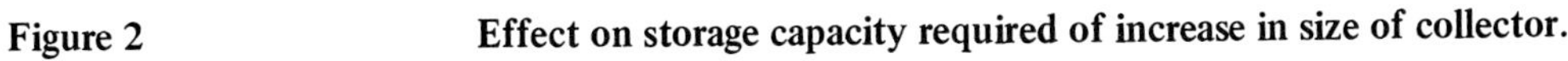

Figure 2 **Effect on storage capacity required of increase in size of collector.**

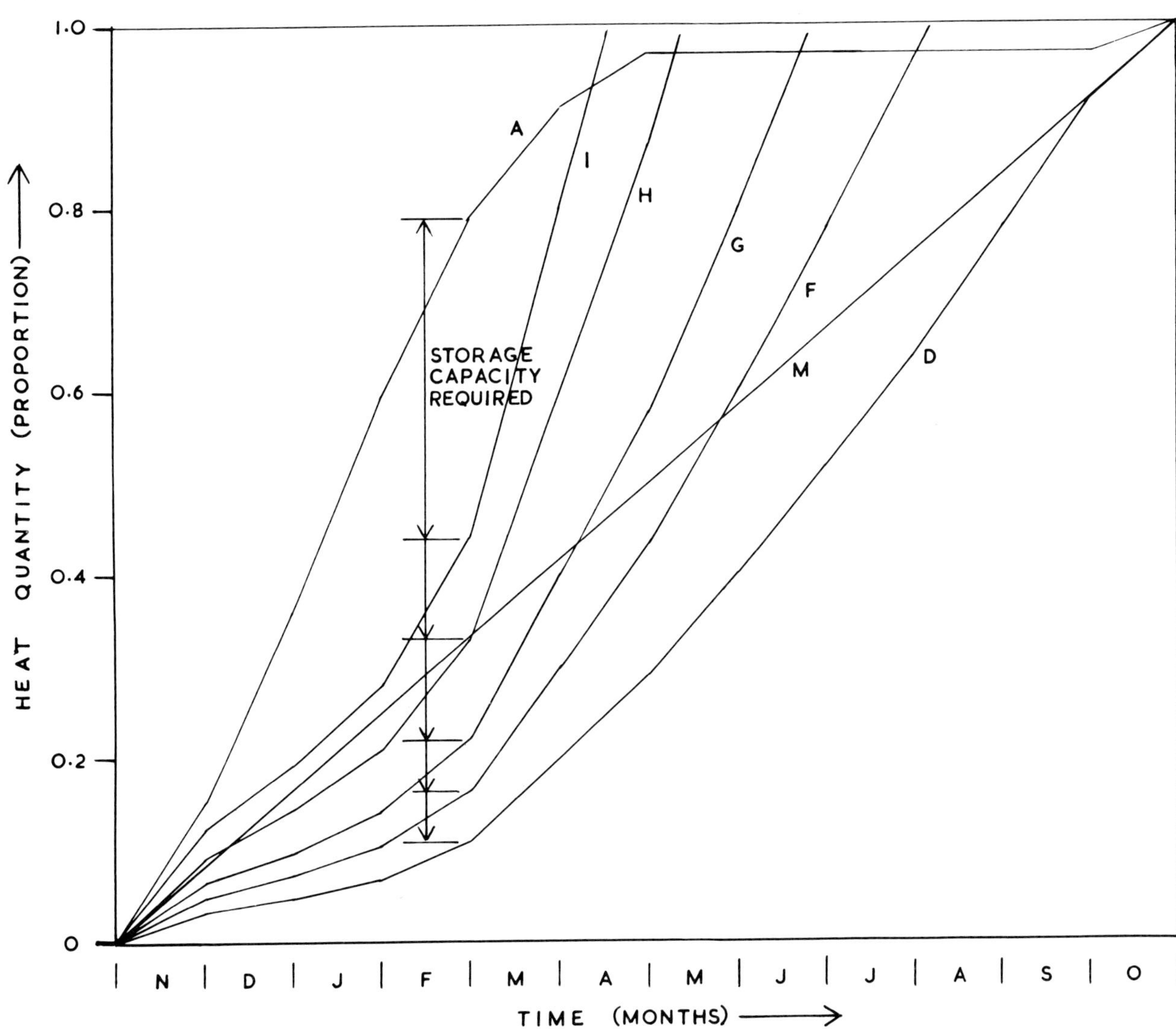

year, about 20% less heat is collected due to the reduction in efficiency of the collector over the winter months and the size of the collector must be increased to compensate for this. The storage capacity required is increased to 71% of the annual heat requirement. If the efficiency of the collector is only 50% of that indicated by curve 'C', at all times, a collector having an area twice as great would be needed but the storage capacity required would be the same.

It will be seen from curve 'D' in Fig 1 that the contribution to the annual heat requirement made by the collector during the period from the end of October to the beginning of March is small, compared with the annual total, and it is tempting to consider collecting heat during the period March to October only, since problems of winter operation will then be obviated. The cumulative total of heat collected is then as indicated by curve 'E'. It will be seen that the storage capacity required is increased to 81% of the annual heat requirement. The collector area would also have to be increased, by 12%. Greater heat collection would be achieved during the summer months by using an inclined collector but the case of a vertical collector has been considered because of its higher collection capability in the winter months.

The size of the solar collector used in conjunction with a storage system can be increased so that the quantity of heat collected over a year exceeds the annual heat requirement. The storage capacity required will then be reduced. The effect of increasing the size of the collector is illustrated in Fig 2, in which the heating load and quantity of heat col-

Figure 3

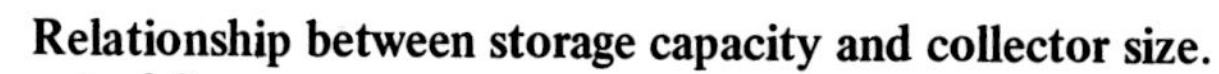

Relationship between storage capacity and collector size.

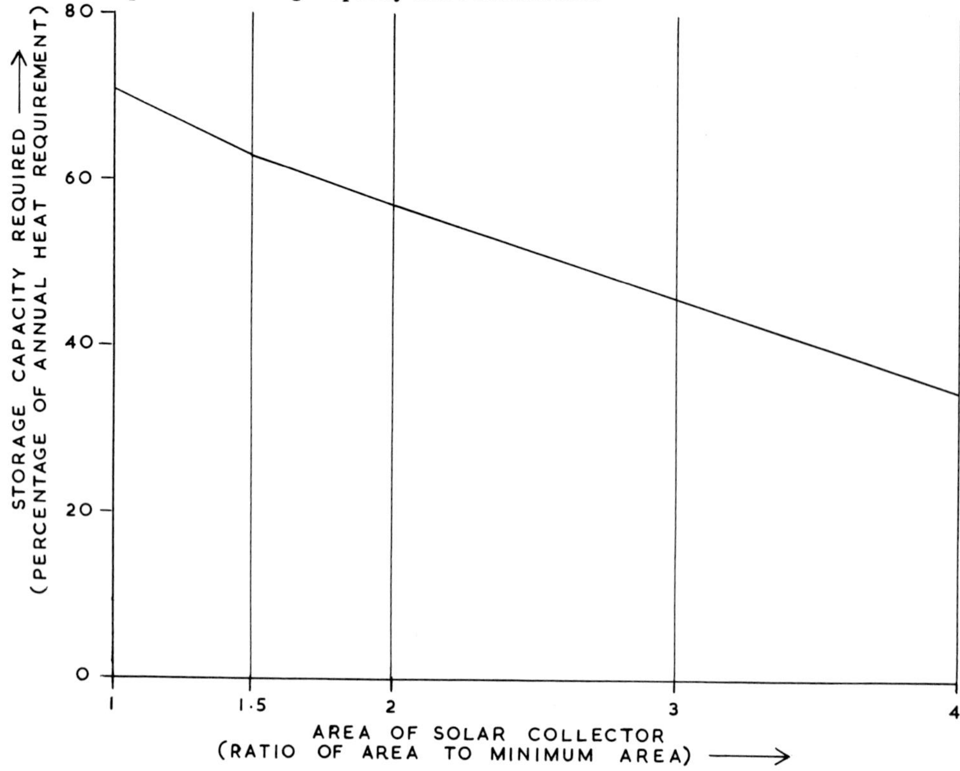

Figure 4

Variation in storage and collection costs.

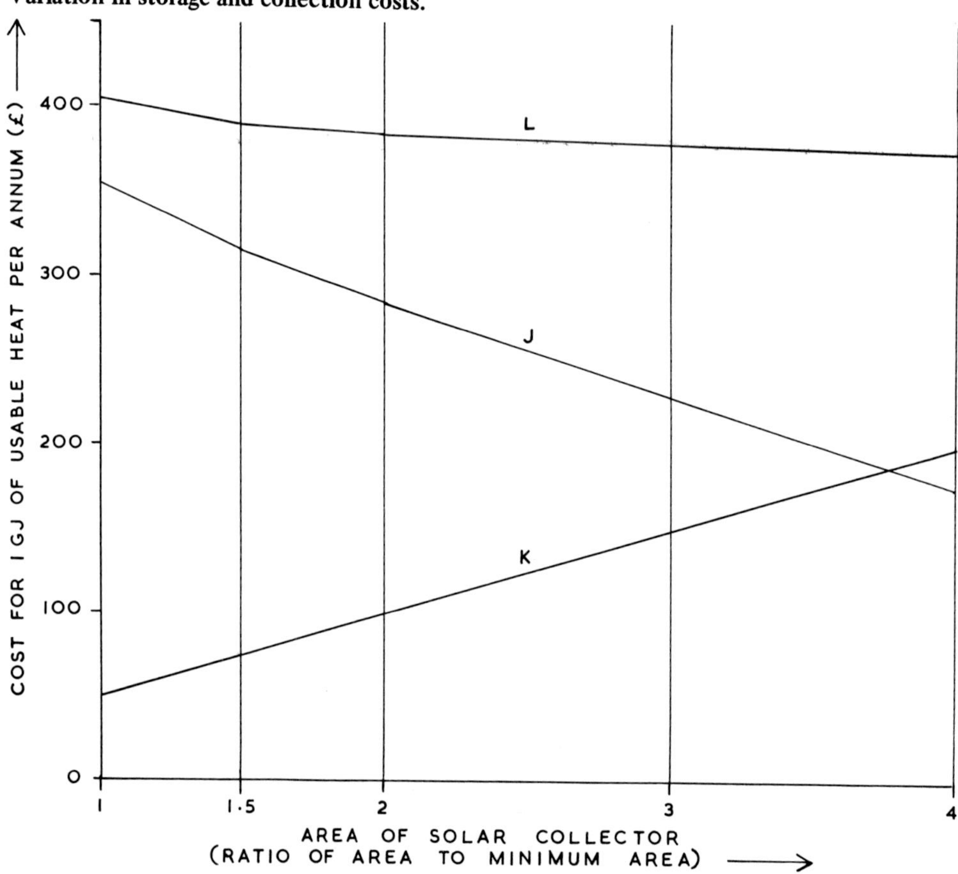

lected are shown cumulatively commencing at the end of October. With the minimum size of collector, the quantity of heat collected exceeds the heating load in October but falls short in November, so the end of October has been taken to be the time at which the cold period of the year commences. As in Fig 1, curve 'A' represents the cumulative heating load and curve 'D' the cumulative quantity of heat picked up by a collector of minimum size having an efficiency over the year as indicated by curve 'C' in Fig 1. The difference between the values indicated by curve 'A' and a cumulative heat collection curve such as curve 'D' represents in Fig 2 the quantity of heat drawn from storage at the time and the maximum difference therefore represents the storage capacity that must be provided.

Curves 'F', 'G', 'H' and 'I' show cumulatively the quantity of heat collected by collectors of 1.5, 2, 3 and 4 times the minimum area required, respectively. The larger the collector, the greater is the rate at which the store is heated during the spring and summer and the earlier is the time in the year when the store becomes fully heated. After this time, indicated by the intersection of the heat collection and heating load curves, it is assumed that any more heat collected before the winter must be rejected unless used directly.

A general analysis of the kind given here, lacking precise values, has limitations for assessing costs but it is instructive, in putting the potential role of heat storage in perspective, to look at the order of costs. A plot of the storage capacities indicated by the curves of Fig 2 is shown in Fig 3. It will be seen that by increasing the collector area to four times the minimum area, the storage capacity required is reduced from 71% to 35% of the annual heat requirement. On the basis of the assumptions made, including average values, an area about twelve times the minimum area would be required to meet the December heating load without long term heat storage.

If it is assumed that the cost of heat storage is £500 per GJ, then the cost of storage is as shown by curve 'J' in Fig 4. If it is assumed that 1 GJ of heat per annum is obtained from an area of one square metre of collector and, further, that the cost of the collector is £50 per square metre, the collector cost is as shown by curve 'K' in Fig 4. With these relative costs, it will be seen that the cost of the collector increases by almost the same amount as the cost of storage falls when the collector size is increased above the minimum and the total cost, shown by curve 'L', varies but little.

Accurate values are clearly required to establish the balance of collector area and storage capacity for minimum cost and a general conclusion is unwarranted. The relative ease with which collection area and storage volume can be provided might be more important than minimum equipment cost in practice. The ratio of ten to one for the cost of storage to that of collection, as defined above, appears to be close to the limit at which inter-seasonal storage will become unattractive on cost alone.

In the case of domestic hot water supply, if the cumulative load is taken to increase linearly over the year it is represented by the straight line 'M' in Fig 2. The conclusion is that about 23% of the annual heat requirement for domestic hot water needs to be stored, if a collector of the minimum size is used and the cumulative total of heat collected is represented by curve 'D'. If no heat is collected during the months November to February inclusive, storage is required to meet the whole demand during these months, one third of the annual total.

Storage with a heat pump

If a heat pump is employed, it will be assumed that the heat source is air, solar energy or both.

There are advantages in providing storage on the hot side of a heat pump, to store the output. The heat pump need then be operated only at times when the weather is favourable and a high coefficient of performance is achievable. Heat can be drawn from the store as required and the heat pump does not need to be sized to meet the peak heating demand. When heat is available at a sufficiently high temperature, typically in warm weather, heat can be taken directly into the store, thereby reducing the power consumption of the heat pump over the year.

Arguments can be advanced for providing storage on the cold side of a heat pump. The heat pump can be operated irrespective of the weather conditions, heat being drawn from the store during unfavourable periods, and, being cold, the store gains heat from its surroundings instead of losing heat. The latter is particularly important when long time periods are involved as even a small rate of heat loss is then significant. A further advantage is that excess heat in the building can be absorbed into the store, whether the heat pump is in operation or not. Cooling of the building can be achieved in this way while meeting the requirement for an input of heat. If extract ventilation is employed, heat can be recovered from the exhaust air, at a high rate continuously, by passing the air through the store before it is discharged to outside.

There is a case for providing two stores with a heat pump, one on the hot side and the other on the cold side. The advantages both of hot-side and cold-side storage are then obtainable and the heat pump can be run either continuously, under substantially constant operating conditions, or intermittently, when desired. Cooling can be obtained without reversal of the heat pump and the maximum rate at which heat can be supplied or absorbed by the system is not dependent upon the rating of the heat pump but only upon the design of the stores. If the heat pump is run all the time, only a small machine is required.

A larger machine need be run only at those times when conditions are favourable, during off-peak periods if mains electricity is used or during favourable weather if wind power is used. Sizing of the two stores will depend upon the objectives, upon whether it is desired to limit either the energy consumption of the heat pump or the capital cost of the system. If the power consumption of the heat pump over a year is required to be low, a large capacity hot store will be required to permit a substantial proportion of the quantity of heat required to be fed into the store directly, during warm weather, instead of via the heat pump. The cold-side storage in relation to the total storage capacity can be increased when the quantity of energy supplied to the heat pump over a year is increased and as the size of the heat pump is increased, the total storage capacity can be reduced, since more heat can then be taken in during the winter.

If power is taken from an external source to operate a heat pump during cold weather, this will contribute to the heating effect and the total storage capacity required will be reduced correspondingly.

Storage temperatures

If a store for space heating is heated by the output of a heat pump, the coefficient of performance of the heat pump will fall as the temperature of the store is increased. If a simple flat plate collector is used to heat the store, directly, the efficiency of the collector will fall as the temperature is raised. Since the heat loss from the store will increase with temperature also, it is apparent that in such cases the storage temperature should be as low as possible. A storage temperature of 40°C is probably close to the maximum that is acceptable with these methods of store heating and, to heat a room directly from the store, the minimum temperature is probably about 25°C.

To use heat stored in this range directly, at a temperature only a little above room temperature, implies the use of a large heating surface. Further, when a heat pump is used or solar energy is collected at typical flat plate collector temperatures, it is apparent that the permissible temperature variation of the store is small. When heat is collected at a higher temperature, by a concentrating collector, or the energy derives from wind power, storage at higher temperatures is possible and this permits a wider variation of the storage temperature.

Unless supplementary heating is provided, domestic hot water supply requires either collection and storage at a relatively high temperature (60°C) or the use of a heat pump to upgrade the heat. In the case of the latter, cold-side storage might be employed with advantage.

For storage on the cold side of a heat pump, the requirements are conflicting. The higher the temperature at which the store is held, the smaller will be the power consumption of the heat pump but the smaller will be the quantity of heat it is possible to collect. Storage temperatures of around 0°C and 20°C are of special interest. With a storage temperature of about 0°C, heat is picked up easily, even from the outside air in this country for much of the winter, but the power required to upgrade the heat is rather high. With an internal heat storage vessel at 20°C, assuming this is the average internal temperature, there will be no uncontrolled heat flow to or from the store and no need to insulate it.

HEAT STORAGE MEDIUM

The traditional medium for heat storage is water. Its cost is minimal, its specific heat capacity is high and it can serve conveniently as a heat transfer medium for transferring heat to and from the store. If the permissible temperature variation is small, though, as when a storage temperature range of 25°C to 40°C is all that is available, the storage capacity provided by sensible heat storage is low, even with a substance such as water having a high specific heat capacity. A small temperature range makes latent heat storage particularly attractive as an alternative and since melting a substance does not pose the containment problems that vaporisation does, substances that melt and solidify within a narrow temperature range (not necessarily at a specific temperature) merit consideration.

When common substances are examined, a particular group that stands out is the salt hydrates, substances which contain water of crystallisation. The heat-storing properties of some salt hydrates, such as Glauber's salt (sodium sulphate decahydrate) and sodium acetate trihydrate have been known for many years and a particular textbook of physical chemistry cites their use "as thermophores or heat-producing mixtures for warming purposes" (1). When heated, the crystals of some of the salt hydrates melt at a convenient temperature and the salt dissolves in the water to form a solution. This is accompanied sometimes by the absorption of a relatively large quantity of heat. On cooling, the hydrate crystallises from the solution with the liberation of heat.

Because of the high water content, the cost of salt hydrates is commonly low. Many of them occupy a smaller volume in the solid state than does the solution from which they form, so there is not necessarily any danger of fracturing a rigid container on solidification, and they do not in general constitute a fire hazard. The latter is particularly important when the use of a large quantity is contemplated in a building. Several salt hydrates, being inorganic and containing a high proportion of water, can be used to improve the fire resistance of materials by their addition.

The use of salt hydrates in heating apparatus was described in a British patent specification published in 1935 (2). Reference is made in the specification to the use of sodium thiosulphate pentahydrate, which melts at 48°C. The use of Glauber's salt (melting point 32°C) for storing solar heat was proposed in the late 1940's (3) and the use of disodium phosphate dodecahydrate (melting point 35°C) for the storage of heat supplied by a heat pump was investigated in the 1950's (4). In considering their use in bulk quantities in buildings, it is necessary to bear in mind the chemical effects of such substances on building materials and the risk of contact. Two other substances which have a melting point close to 30°C, and are of low cost, are calcium chloride hexahydrate and washing soda (sodium carbonate decahydrate). Prospectively, washing soda (melting point 32.5°C) is capable of storing about 350 MJ per cubic metre at its melting point. Ways must be found though to realise the full potential of the substances. Interest in the salt hydrates for heat storage has intensified in recent years and their use has been discussed in recent articles (5, 6).

At a relatively high temperature level, the author has found that trisodium phosphate dodecahydrate, with water added (7), will store 550 MJ per cubic metre over the temperature range 40°C to 95°C, about 2.4 times the storage capacity of water.

For storing heat at room temperature, sodium chromate decahydrate (melting point 20°C) can be used, borax being added to promote crystallisation, but sodium chromate solutions are dermatitic. For heat storage within but over the temperature range of roughly 10°C to 30°C some encouraging results have been achieved by the author with mixtures of washing soda and water glass (sodium silicate solution).

For the storage of heat at low temperature, around 0°C, ice can be used. Not only is

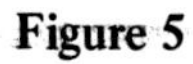

Figure 5 Basic arrangements for heat storage unit.

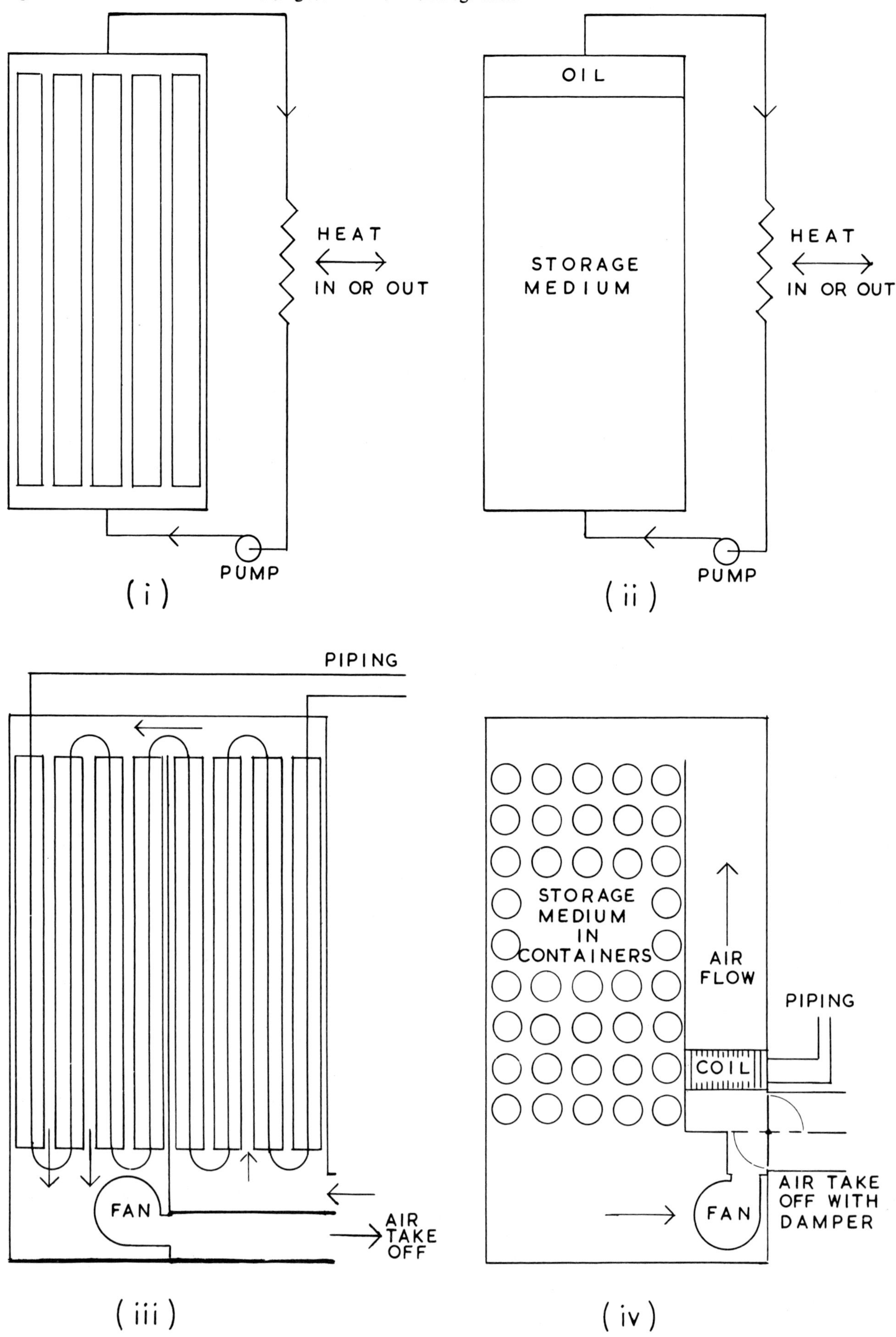

its cost minimal but its latent heat of fusion is high (333 kJ/kg) and it will store 305 MJ per cubic metre in melting. The merits of ice for storage have been recognised since the earliest days of air conditioning and its use in storage heating equipment has been described (8). Problems of containment and evaporation can be obviated by sealing the water in a container. Fracture of the container on freezing of the water can be prevented by allowing a free space above the liquid and adding a small quantity of, for example, ethylene glycol or an appropriate salt. Providing the amount of ethylene glycol, or salt, added is small, freezing of the bulk of the water fairly close to 0°C can be accomplished but the mass will not freeze solid until cooled to the eutectic temperature. Expansion occurs during solidification but is accommodated by displacement into the free space.

DESIGN OF HEAT STORES

When a liquid such as water is used as the heat storage medium it is convenient to circulate the water through a heat exchanger for transferring heat to or from the store, but if latent heat storage is used it may not be possible to circulate the heat storage medium itself. If solidification of the medium occurs, as with a salt hydrate, heat must be conducted through the medium and the equipment must be designed accordingly. The provision of a large area for heat flow will avoid excessive temperature gradients being set up. It needs to be borne in mind that with long term heat storage, heat flows are small in relation to the volume of the storage medium and problems of heat transfer will not necessarily arise. Heat transfer might be quite adequate with a simple tank, perhaps of narrow section, having a heat exchanger contiguous with or embedded in the walls.

Some basic arrangements for a storage unit which can provide good heat transfer are shown in Fig 5, but many variations are possible. One technique is to use an intermediate liquid heat transfer medium such as oil or water. A storage vessel can be filled with the liquid and containers holding the storage medium immersed in it, as shown in Fig 5(i). By using cylindrical containers, free circulation of liquid between them is ensured and good heat transfer can be achieved between the liquid and the bulk of the heat storage medium if the containers are relatively small in diameter. The liquid can be circulated through any number of heat exchangers for transferring heat to or from the store. The evaporator or condenser of a heat pump can be immersed in the liquid. Another arrangement, in which oil is used as a heat transfer liquid, is shown in Fig 5(ii). In this case, the oil is pumped through the storage medium, agitating it and providing very good thermal contact. The use of this latter technique with disodium phosphate dodecahydrate has been investigated (4).

Air can be circulated through a store instead of a liquid. This will increase the size of the store but has practical advantages. The storage vessel does not need to be watertight and corrosion problems might be eliminated. Air may be used to transfer heat either to or from the store, or for both. Unless air is used to perform both functions, a second means of transferring heat will be required and this can be a pipe coil in the storage medium, as illustrated in Fig 5(iii). If the storage medium is ice, for example, the pipe coil can be the evaporator of a vapour compression heat pump. If the output of a heat pump is stored, the pipe coil can be a condenser. Water heated by a solar collector could be circulated through the pipe coil to heat the store. The air is circulated over the large external surface area of the containers holding the storage medium.

When air is used to transfer heat both to and from the storage medium, the air may be simply recirculated within the storage vessel, as shown in Fig 5(iv), in which case transfer of heat to or from the coil shown is controlled by operation of the fan. The air can instead be circulated through a solar collector for heating the store or supplied to a room for space heating, or cooling, according to the temperature. If air is circulated through the storage vessel for space heating at times when heat is available from the coil, either the air can be passed over the coil first, so as not to deplete the store unnecessarily, or the coil can be used to boost the temperature of the air from the store. The use of air as the heat transfer medium requires the provision of a very large surface area for heat transfer when the temperature differences available are small to avoid excessive fan power consumption.

Whilst the characteristics of the different arrangements possible are important, of perhaps more importance in assessing the feasibility of latent heat storage systems are the practical questions of the physical handling of the storage medium and the cost of containment. It is feasible to think in terms of filling a pit with a molten salt hydrate, delivered in bulk, providing spillage would not have deleterious effects or could be eliminated and it might be that a pipe coil embedded in the walls would be adequate for heat transfer on a long time scale.

The traditional approach to the design of storage heating equipment in this country though has been to make it possible, for ease in assembling or dismantling the equipment, to provide the storage medium in quantities sufficiently small as to be handled manually with comfort, typically in increments of perhaps 5 kg. A method of achieving this is to contain the storage medium in ordinary thin-walled glass bottles. This not only provides a very cheap form of containment but also a very large surface area for the transfer of heat to or from air with small temperature differences. The size of bottle chosen determines the ratio of surface area to volume but this will be more than adequate for the purpose with any practical size. Further advantages of glass bottles are their rigidity and corrosion resistance. They can also be filled rapidly and economically using existing equipment.

When ice is used, it can be treated as a special case of latent heat storage. It can be contained in bottles, using the technique previously described, but is probably as conveniently contained in metal or plastic tanks. The difference is that in the case of water the tank can be filled and drained off if necessary, without undue difficulty, on site, so the same problems do not arise as with the use of salt hydrates. An evaporator can be immersed in the water for freezing it, or coiled on the tank. Natural or forced circulation of air over the walls of the tank or through tubes in it will permit heat to be taken in at a high rate, unless the air is at a very low temperature.

If heat is stored at an elevated temperature, the store is preferably located inside the building heated, so that the heat emission from the store will contribute fully to heating the building. A low temperature store, at around 0°C, can with advantage be located outside the building. Not only will the store then not take heat from the building in winter, but a substantial quantity of heat can be picked up from outside during the warmer periods of the winter, without the supply of fan power. Natural heat gains to the store in winter can be increased by exposing a storage vessel at low temperature to solar radiation. This can be achieved easily if the vessel is mounted in a roof space.

APPLICATION

The application of long term heat storage will depend principally on the cost and space requirements. The problem of relating the apparent material cost to the likely actual cost of equipment makes the validity of general cost comparisons doubtful but, as a first step, it is useful to examine the order of material costs since this gives some idea of the economic limits. Taken with the more definite space requirements, this enables an assessment of the potential role of heat storage to be made.

If heat is stored in the temperature range 25°C to 40°C, to permit it to be used directly for space heating, the material cost using the cheaper salt hydrates, contained in bottles of minimal cost, is not likely to be less than £300 per GJ, so it is unlikely that the actual cost of heat storage using these materials will be much less than the figure of £500 per GJ taken previously. For buildings of traditional design, with which the annual space heating requirement of a house might be in excess of 40 GJ, the cost of inter-seasonal storage using these materials is probably prohibitive.

If the annual space heating requirement of a house is reduced, by the use of insulation, to say 18 GJ per annum and 71% of the annual heat requirement for space heating is stored, the cost of the heat storage material, in containers, is still not likely to be less than £4000, for space heating alone, and the cost of the equipment will be higher.

The maximum energy storage density likely to be achieved with salt hydrates, in this

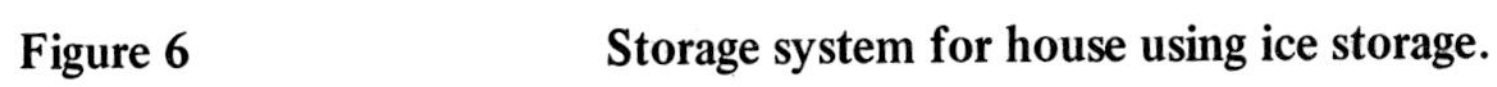

Figure 6 **Storage system for house using ice storage.**

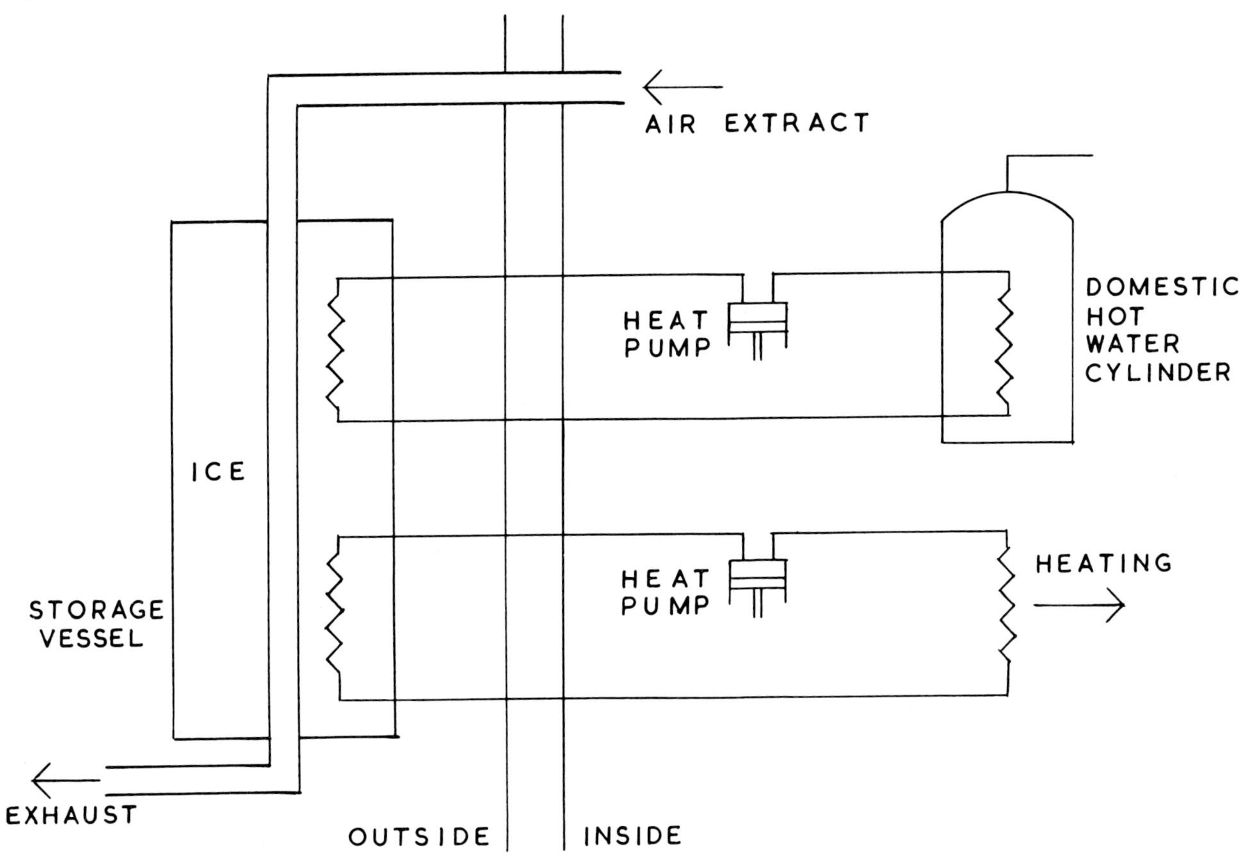

temperature range, is of the order of 400 MJ per cubic metre, so that the net volume of the storage medium would be somewhat greater than 30 cubic metres. The overall volume of the store might be 50 cubic metres, perhaps 25% of the volume of the accommodation provided, and the weight around 50 tonnes. For comparison, assuming a working temperature range of 15 K (25°C to 40°C), a net volume of water of 205 cubic metres would be required, over 200 tonnes, for the same heat storage capacity. To store the same quantity of heat, with the same temperature range, about 1000 tonnes of concrete would be required, a volume of about 400 cubic metres.

If a heat pump is employed with only cold-side storage, using ice, the cost of the equipment will be much lower than if heat is stored at a temperature at which it can be used directly for space heating. With the use of ice as the storage medium, it is convenient to collect heat from air exhausted from the building and the need for external equipment is obviated, unless a windmill is used to provide the power for the heat pump or it is decided to mount the storage vessel outside. Heat can be taken from a single storage vessel if desired for both space and domestic hot water heating and such an arrangement, shown in Fig 6, will be assessed by taking the example of a well-insulated house. It will be assumed that the annual heat requirement for space heating is 18 GJ and, additionally, 12 GJ per annum is supplied for heating domestic hot water.

The cumulative space heating load is represented in Fig 7 by curve 'N', derived from curve 'A' of Fig 1. The domestic hot water load is assumed constant over the year, giving a total load, for space and water heating, as indicated by curve 'O' in Fig 7. If a heat pump for domestic hot water supply raises the temperature of the water from 10°C to 40°C and supplementary heating then raises the temperature of the water to 60°C, over a year 4.8 GJ of supplementary heat will be supplied. Curve 'P' indicates the total load less the supplementary heat supplied and curve 'Q' in Fig 7 the quantity of heat drawn from storage by heat pumps for space and water heating on the assumption that each has a coefficient of performance of two.

Figure 7 **Cumulative heat collection and load curves for system shown in Fig 6.**

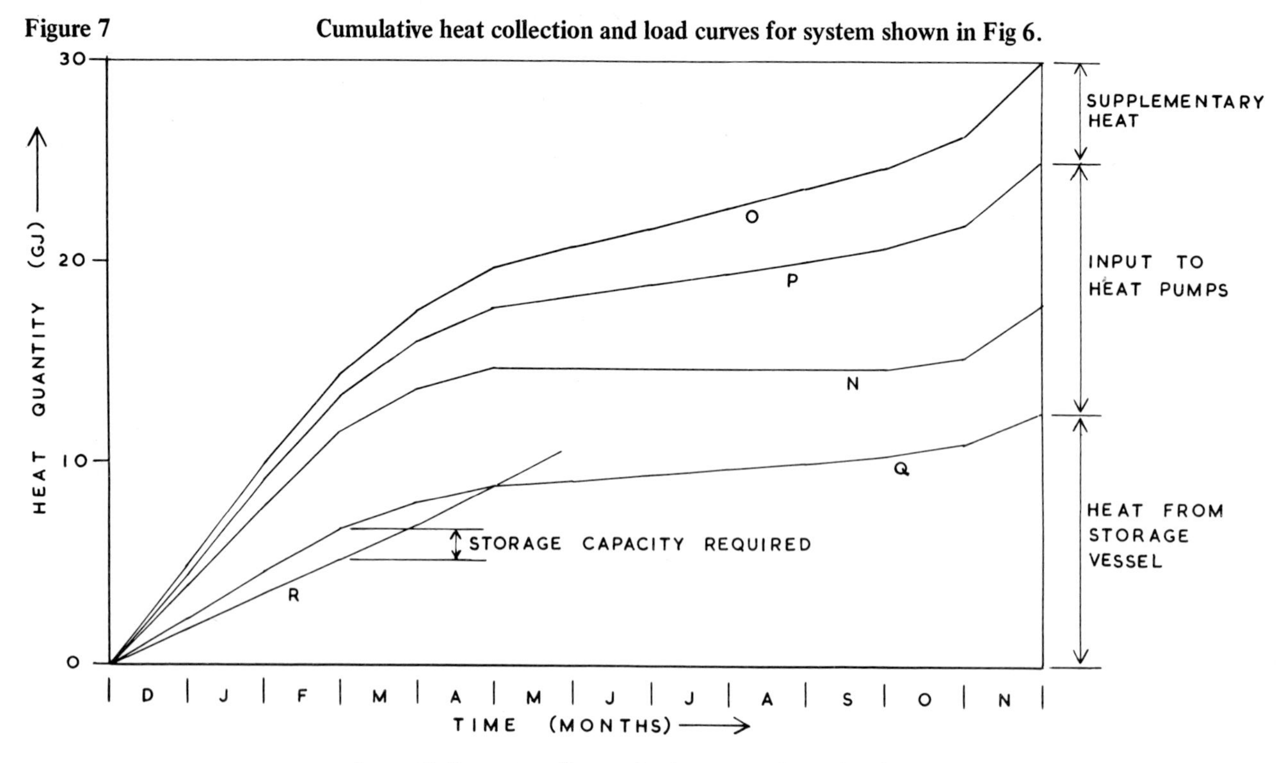

A ventilation rate of one air change per hour has been assumed to occur in the house and the extract ventilation system will be taken to exhaust a volume equal to 0.5 air changes per hour continuously. The air might be extracted from the kitchen and bathroom to remove moisture. In the arrangement shown in Fig 6 the air is discharged to outside through the storage vessel but it could equally well be passed over the walls of the vessel. The vessel will be treated as a cooler at 0°C having a contact factor of 0.8. If a modest allowance is made for moisture pick up by the air in the house (0.1 kg/h), mean dew point values for outside air indicate that, after the air has been heated by passage through the dwelling, the quantity of heat supplied to the storage vessel by the exhaust air will be as shown by curve 'R' in Fig 7.

During the period from March to November, inclusive, the heat input to the storage vessel from the exhaust air exceeds the quantity of heat withdrawn by the heat pumps.

Figure 8 **Heating and cooling system using ice storage.**

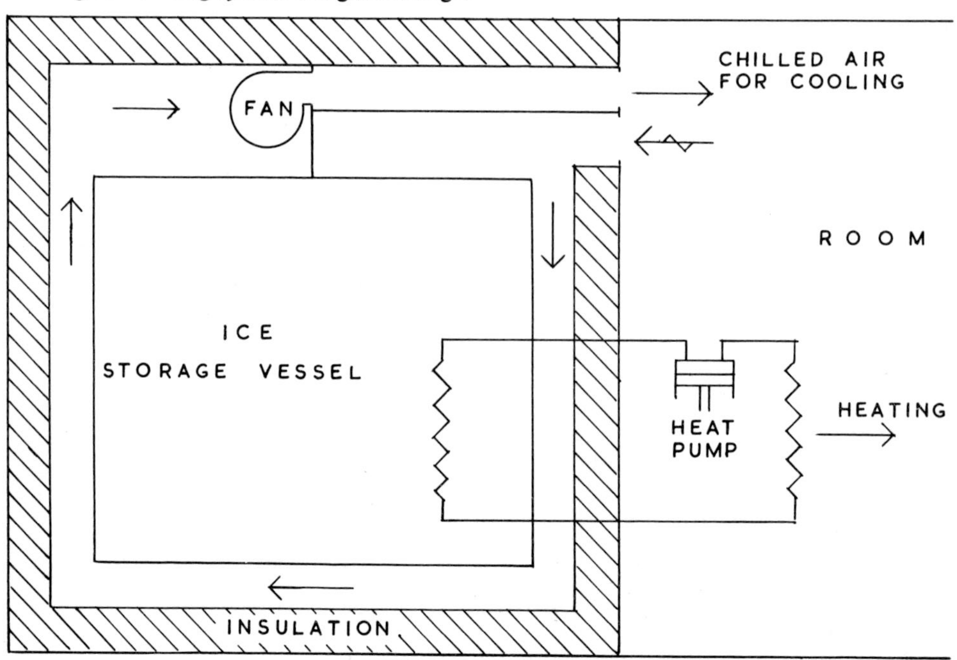

Figure 9 **Cumulative heating and cooling load curves for system shown in Fig 8.**

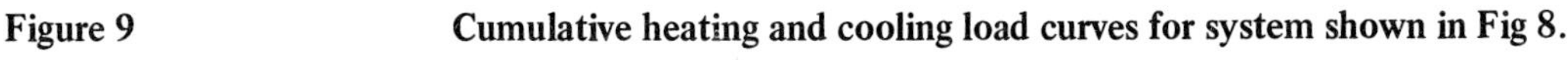

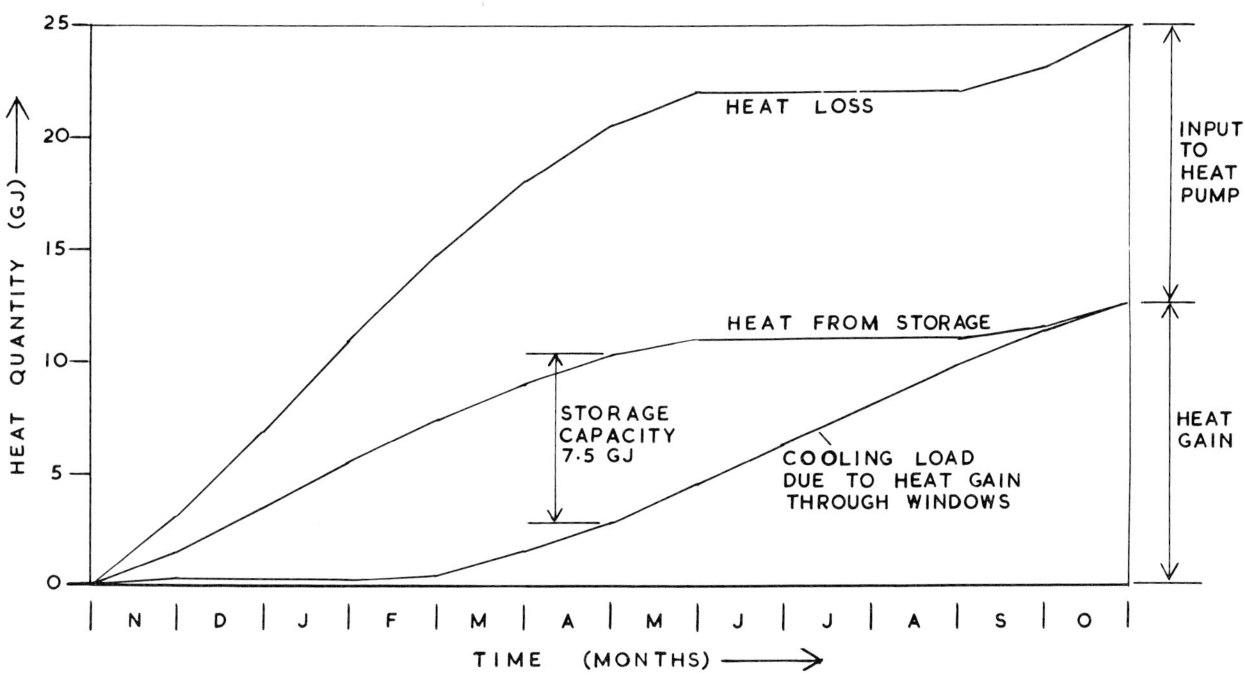

Hence, the cumulative heat quantities have been taken, in drawing Fig 7, from the end of November. The storage capacity required is the amount by which the quantity of heat extracted from the store exceeds the quantity of heat supplied at the end of February, 1.5 GJ, 5% of the total annual heat requirement. The volume of ice required to store this amount of heat is 4.9 cubic metres, a weight of 4.5 tonnes, and it is reasonable to suppose that a space allocation of 8 cubic metres would be adequate for the store, perhaps 4% of the volume of the accommodation in the house.

The cost of the storage vessel and associated work is not likely to be prohibitive and would be small compared with the cost of inter-seasonal storage using a salt hydrate. It will be seen from Fig 7 that 17.4 GJ per annum, 58% of the annual heat requirement is supplied to the heat pumps and in supplementary heating of domestic hot water over a year and that 12.6 GJ per annum, 42% of the heat requirement, is collected by the store from the exhaust air. In practice, for much of the time the temperature of the store will be above 0°C and a higher coefficient of performance would then be expected from the heat pumps.

Variations on the simple system shown in Fig 6 might be worthwhile. If outside air is allowed to circulate over the storage vessel, so that heat is picked up from outside during warmer periods of the winter, the storage capacity required will be reduced. In the summer, air might be ducted into the building over or through the storage vessel to provide cooling. If summer cooling is required, there is a case for not exhausting air through the storage vessel after the end of the winter but for holding the ice formed during the winter until summer cooling is required and then using it for refrigeration. An ice storage system has the capability to remove heat very rapidly from a room and large heat gains, such as those due to solar radiation through windows, can be accepted by the store.

The summer refrigeration potential of a system that uses ice as the storage medium for winter heating might be exploited in a commercial building, such as an office, or in the case of a school. To maximise the summer refrigeration potential, heat would not be taken in at all in winter, but the maximum quantity of ice would be formed during the winter months. A simple arrangement is shown in Fig 8. Cumulative heating and cooling load curves, over a year, are shown in Fig 9, for a rectangular room having a design heat load of 5.5 kW, a floor area of 100 square metres, a ceiling height of 3 metres and a window area equal to 15% of the floor area.

A ventilation rate of 2 air changes per hour and use for 12 hours per day for 5 days per week have been assumed in drawing the curves. Miscellaneous heat gains have been ignored but would contribute to heating in practice. The windows have been assumed to face South, to be double glazed and to be fitted with an internal blind. The cumulative load curves indicate that a storage capacity of 7.5 GJ is required, 60% of the annual cooling requirement, 30% of the annual heating requirement. A volume of ice of 25 cubic metres would be required, just over 8% of the volume of the room, so a space allocation of 15% of the volume of the room might be required for the store.

The coefficient of performance of the heat pump has again been assumed to be two and extraneous heat transfer has been ignored. Only fan power would be required for summer cooling and the annual energy input to the heat pump is that required for winter heating. In effect, the windows perform the function of a solar collector for winter heating and winter operation of the heat pump provides summer refrigeration. At the end of a cool summer, heat could be taken in to melt any residual ice by circulating outside air over the storage vessel or room air during periods when the room is unoccupied.

CONCLUSION

Since detailed analysis is necessary to establish the economics of long term heat storage, this can only be done in a specific case. Some tentative conclusions, of a general nature, can be drawn, however, from the sizes and costs indicated for the hypothetical cases considered.

It is clearly technically feasible to collect heat throughout the year to meet the heating demands of winter by the use of storage. The development of techniques using salt hydrates is likely to permit this to be done without the use of a heat pump, but at high capital cost.

Ice provides a low cost storage technique but depends upon the use of a heat pump to extract heat from the store. A power input to the heat pump equal to around 50% of the heating requirement is perhaps necessary and the source of this power is a crucial consideration.

If both cold-side and hot-side storage are used with a heat pump, the relative importance of minimising capital cost or power consumption is likely to determine the division of storage capacity between the two, but the provision of summer cooling might also be an important factor. High storage capacity on the hot side will permit very low annual power consumptions to be achieved, since a large quantity of heat can then be supplied directly to the hot store, when the weather conditions are favourable. High storage capacity on the cold side will not only be much cheaper to provide but might allow summer cooling at little additional cost, in energy or money.

If ice is used for storage for summertime refrigeration as well as winter heating, a highly insulated building is probably not essential. A well insulated building is desirable though with long term storage to limit the storage capacity required, the power consumption of a heat pump or both.

REFERENCES

1. Taylor, H S, and Taylor, H A, *"Elementary Physical Chemistry"*, Third Edition, 1942, p 300. Van Nostrand.
2. British Patent Specification, No 471 505, 1935.
3. Telkes, M, *"Solar House Heating – A Problem of Heat Storage"*, Heating and Ventilating, May, 1947, 68-75.
4. Etherington, T L, *"A Dynamic Heat Storage System"*, ASHAE Trans, **64**, 1958.
5. Telkes, M, *"Solar Energy Storage"*, ASHRAE Journal, Sept 1974, 38-44.
6. Pillai, K K, and Brinkworth, B J, *"The Storage of Low Grade Thermal Energy Using Phase Change Materials"*, Applied Energy, July 1976, 205-216.
7. British Patent Specification, No 1 396 292, 1975.
8. British Patent Specification, No 1 396 293, 1975.

DISCUSSION

Professor B. J. Brinkworth (University College Cardiff)

I have to admit to a certain unease about some of Dr Randell's propositions and I would like to highlight two; on his strategy for finding the storage requirements for buildings and the materials he has used for storage.

Concerning the strategy, it seemed that the treatment in the paper is sufficiently simplistic to be very misleading because it deals only with the aggregate energy transfers and makes no reference to temperature variations. Energy can be put into a store only if the collector temperature exceeds the store temperature and as the storage temperature changes, the efficiency of the collector will change. Dr Randell gave average collector efficiencies to arrive at some of his results but those averages cannot be given without making an elaborate system study. Such a study would indicate efficiencies to be very much lower than the figures quoted by Dr Randell. The use of what I think are rather fallacious figures has led Dr Randell to come to some conclusions which might be questioned. An example is the statement that the collector area could be increased if the system is not satisfying the demand, and that a collection area about twelve times the minimum value that was established would be required to meet the heating load in December, without any long term heat storage at all. In fact, when temperatures are taken into account it is found that for all practical purposes it is not possible to meet the December loading with any collection system whatever.

Then the next point is on the materials used for energy storage. Dr Randell has shown examples of hydrated salts, I wondered why he had given particular attention to this group of storage materials. These have a long history in the energy storage business but it seems that everyone who has worked with salts has finally abandoned them largely because of two problems. One of these was the separation of the phases that most hydrated salts exhibit on repeated thermal cycling. The salt used most because of its easy availability and cheapness is Glauber's Salt (sodium sulphate decahydrate). Although a solid at room temperature it is a very peculiar substance, which on heating goes into four different phases at different temperatures. The first change occurs at 32°C, a convenient temperature for heat storage. The salt does not actually melt at this temperature but some of the sodium sulphate goes into solution in the water of crystallisation, depositing the rest as an anhydrous solid, which having a high density, sinks to the bottom. A similar process seemed to have occurred in one of the sample bottles shown by Dr. Randell. On repeatedly cycling such media, the separation of phases results in an energy storage density only 16% of what would be expected if the full phase change were taking place. This problem probably applies to most hydrated salts. None of those known to me really melt in the congruous sense that many of the more familiar solids do. Have these media anything to recommend them other than their very low cost?

Dr. J. E. Randell

Any efficiency curve over the year can be substituted for curve C in Fig 1 of Chapter 12 and the corresponding heat collection curve found, but the relative closeness of the heat collection curves shown in the figure indicates that lack of exactness in the shape of the efficiency curve does not invalidate a general conclusion. The area of collector required will be determined by the efficiency curve but I think the analysis given provides a reasonable indication of the storage requirement. My intention was only to indicate roughly the storage capacity required and the analysis is based on average values. The relative size of the collector needed for heating in December is based on an average value and was given purely for purposes of comparison. With less than average receipt of radiation in the month of December in a particular year, the quantity of heat collected would be inadequate. The quantity of heat collected in a particular month does not have the same significance when heat storage requirements are assessed on the basis of collection over a year as when individual months are considered independently.

Salt hydrates are not only relatively cheap but, being perhaps nearly two-thirds water, they resist fire. I do not favour the use of organic substances in quantity because of the fire hazard. Salt hydrates proposed for heat storage are also non-toxic. In the case of Glauber's Salt, a soluble sulphate, I am concerned about possible attack of concrete. The effect of phosphates on building materials similarly is a matter of concern to me. Washing soda is a salt hydrate which, if it can be used, does not raise the same question in application. There are fundamental difficulties in using washing soda but, having experimented with it for some years, I remain optimistic on finding a satisfactory formulation for its use. In tests on the use of trisodium phosphate dodecahydrate for heat storage, a mix has been cycled over a full heating season and no loss of storage capacity was observed after a year. Sodium chromate decahydrate, with borax added, can be cycled repeatedly. This salt dissolves completely in its water of crystallisation. I do not advocate its use, because it is dermatitic, but success with sodium chromate decahydrate encourages me to pursue the use of washing soda.

Dr. R. W. Todd (National Centre of Alternative Technology)

Regarding heating in mid-winter, it is my experience based on design calculations for the building at Machynlleth that even to heat the building in an average December, without storing heat from previous months, we would need something like 1000 m^2 of double glazed collector rather than the 100 m^2 we installed. Has Dr Randell taken into account heat loss from the store? In trying to design the Machynlleth building this seemed an important factor. With a small collector and a large store, there is the prospect of holding heat for very long periods without significant amounts of top up later in the year. This has expensive repercussions on the tank insulation; Mr Seymour-Walker mentioned that the BRE house has a volume of polystyrene equal to the volume of water. We were a little frightened of the escalating cost of the polystyrene and this pushed us towards a larger collector, enabling the store temperature to be maintained until fairly late in the year and requiring a run down time for the store of about two months only.

Dr. J. E. Randell

If the heat store is located within a building, so that heat emission from the store contributes to the heating, and the storage temperature is kept to around 30°C, which is not much higher than the temperature of a floor heating panel with embedded pipes, I think long term storage of heat is feasible. Perhaps the greatest advantage though of using ice for heat storage is that the store gains heat instead of losing it. By locating an ice storage vessel outside, perhaps 50% of the heat required to melt the ice can be picked up from the outside air.

J. D. L. Harrison (AERE Harwell)

I did not understand Dr Randell's advocacy of ice as a storage medium for use on the low temperature side of a heat pump. Since the ambient atmosphere is normally significantly above 0°C, the heat pump should perform better if the ambient air is used as the heat source rather than a tank of ice water mixture at 0°C. Perhaps he could clarify this?

Dr. J. E. Randell

Air at a higher temperature is theoretically a more attractive heat source than ice. In practice, though, because of the large temperature difference at an outside air heat exchanger, the evaporation temperature with a conventional vapour compression heat pump taking heat from the outside air might be -15°C to -10°C. With ice storage the bulk of the water can be frozen with an evaporation temperature of around -5°C to -10°C, so the evaporation temperature when ice storage is used is not necessarily lower than with a conventional heat pump. It can be argued that with a conventional heat pump the evaporation temperature will be higher during mild weather, but there are practical advantages in using ice storage. An ordinary refrigeration system can be used, it operates under substantially constant conditions, no control problems arise and the system need not be reversed for cooling. The system can be run at times when off-peak electricity is available and no defrosting is necessary. Ice formation on the outside of a storage tank is unimportant.

Dr. G. S. Saluja (Robert Gordon's Institute of Technology)

Using ice as cold storage the coefficient of performance of the heat pump would be very low since the storage temperature may easily fall to -10°C. It will be even more difficult to justify the use of ice as storage medium for summer cooling, where the low side could be operated at a much higher temperature. By using ice the fall in coefficient of performance would be very significant.

Dr. J. E. Randell

A coefficient of performance of two was assumed in the paper as a conservative value. In practice, the water in an ice storage vessel will be above 0°C for much of the year and the coefficient of performance should be somewhat higher. To form ice for summer cooling is not a departure from traditional air-conditioning practice. There is a penalty in using ice storage in that the refrigeration system is operated at a lower temperature than is strictly necessary for cooling, but I think the price is a small one for the benefit obtained.

In one major building in London an ice collar is maintained on the evaporator coil of the refrigeration plant, as a means of load smoothing, and my argument suggests nothing different. Less power is consumed by operating at a higher temperature, but with ice storage all the heat rejected is available for winter heating. Heat can be taken into an ice store at a very high rate and large solar heat gains can therefore be accepted. This permits windows to be used effectively as solar collectors. The argument that window areas should be restricted to limit solar heat gains is not necessarily applicable when ice storage is used. If a larger glazed area were employed in the example illustrated in Fig 8 of Chapter 12, the system would still be able to cope with the peak heat gain, but the annual energy requirement would be increased.

13. Ambient Energy—Criteria for Building Design

P. Burberry

BUILDING DESIGN CRITERIA

Many of the methods for extracting energy from the environment are designed to respond only to the particular phenomenon which concerns them and even to this phenomenon only when it exists in a degree favourable to energy extraction. Buildings, however, are inevitably subject to a very wide range of influences and must be designed to give a satisfactory performance over the whole range of variation of the many phenomena concerned and, at the same time, to establish a balance between the often conflicting demands of these phenomena. The very difficult problem of design which this situation poses is made yet more complex since the other functional and economic requirements which the building must meet will very often lead to design solutions which are different to those which would best meet the energy needs and a balance has to be made in the final solution.

If this were a new problem awaiting solution it would present a daunting prospect. Any viable solution would necessarily be regarded as a remarkable achievement. We are, however, familiar with current solutions and, as a result, enjoy no sense of wonder at the sophisticated design achievements that have been made.

In this we are perhaps unfortunate to live in a period when energy from fossil fuels has been cheaply available for twenty five years and, for this period most building designers have been able to work ignoring the external conditions so far as the basic plan, shape and materials of the building are concerned. It is important to remember, however that while progress is not uniformly favourable and, in this context the recent past has seen many backward steps, the real history of architecture is nothing to do with visual styles. It is the record over some three millenia of progressive development of effective functional and environmental design.

At present we may be on the threshold of the most significant new development in building that has ever taken place. Throughout the whole of recorded history the basic form of buildings has been governed not by aesthetics, style, structural or economic considerations but by the necessity to provide natural light and ventilation. The development of electric fans and fluorescent lighting have overcome this constraint. In the past buildings had to conform to the limits imposed by natural light and ventilation. At present there is a technological choice. An entirely artificial environment can be provided if it is desired. Whether such a decision would be correct is not our present concern. The technological possibility is real. Whether the choice remains in the future will depend on the availability of the right forms of energy.

The aspects of the natural environment involved in a consideration of ambient energy and the design of buildings can be grouped into the categories of heat, light, air and moisture movement. It is also necessary to take sound and air pollution into account. Fig 1 summarises these influences. In buildings it is not only the environmental control installations which must be considered in relation to the external conditions. The basic shape may be governed by considerations of lighting, air movement and heating. The materials of construction may be dictated by thermal, moisture and sound considerations. Decisions

Figure 1 **Ambient energy factors in the external environment and their relationship to buildings.**

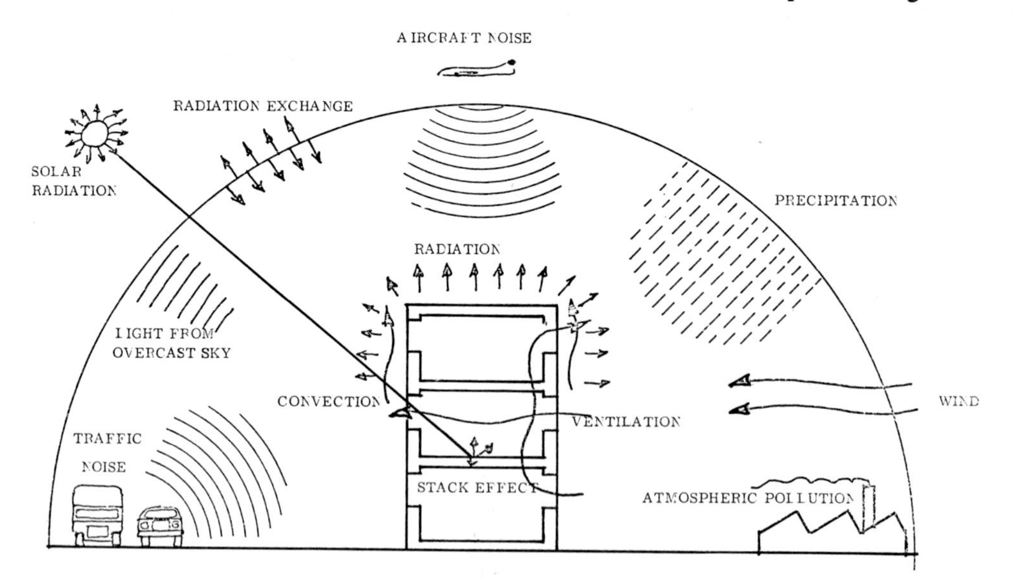

about the windows involve all the factors identified. These considerations of form, fabric and fenestration are the major elements which govern the performance of buildings in relation to the ambient environment and dictate the degree to which the energy there is utilised, or to which additional energy input will be required. The range of variation is considerable. It is clearly possible to design buildings which are economical of energy in winter and comfortable in even the hottest of English summers, but it is also possible to produce buildings which will have the opposite characteristics and may become intolerably hot in summer.

TRADITIONAL BUILDING DESIGN AND AMBIENT ENERGY

The development of central heating, mechanical ventilation and refrigeration have made it possible to maintain standards of comfort in buildings without the need to consider external influences in the basic design. This has resulted from the very recent availability of cheap energy. There is a dramatic contrast with the whole of previously recorded history where wood or peat had to be cut by hand and transported by horse and cart. There was no means of achieving cooling. It is difficult to make a cost comparison with present fuels since the more traditional forms were gathered rather than bought. It is clear, however, that fuel was, in the past, a rare commodity to be conserved rather than the freely available material of the recent past.

It may be thought that there would be valuable lessons to be learnt from historical examples and this is indeed the case. It is possible to observe, in many different climates and many different periods of history, building designs which achieve remarkable results as a result of ways in which they make effective use of available energy or by means of variations of the form, fabric and fenestration ameliorate the effects of unsatisfactory external energy levels. Four examples can be used to show the effectiveness of the traditional approaches.

Example 1 – Utilisation of solar energy

The following quotation is taken from Memorabilia by Xenophon which was written over 2000 years ago. "In houses with a south aspect, the sun's rays penetrate into the porticoes in winter, but in summer the path of the sun is right over our heads and above the roof, so that there is shade. If then, this is the best arrangement we should build the south side loftier to get the winter sun and the north side lower to keep out the cold winds."

The point is amplified in his Oeconomicos. "I showed her decorated living rooms for the family that are cool in summer and warm in winter. I showed her that the whole house fronts south, so that it was obvious that it is sunny in winter and shady in summer."

Figure 2

The method of achieving protection from solar gain in summer while retaining solar gain in winter (for the Northern Hemisphere).

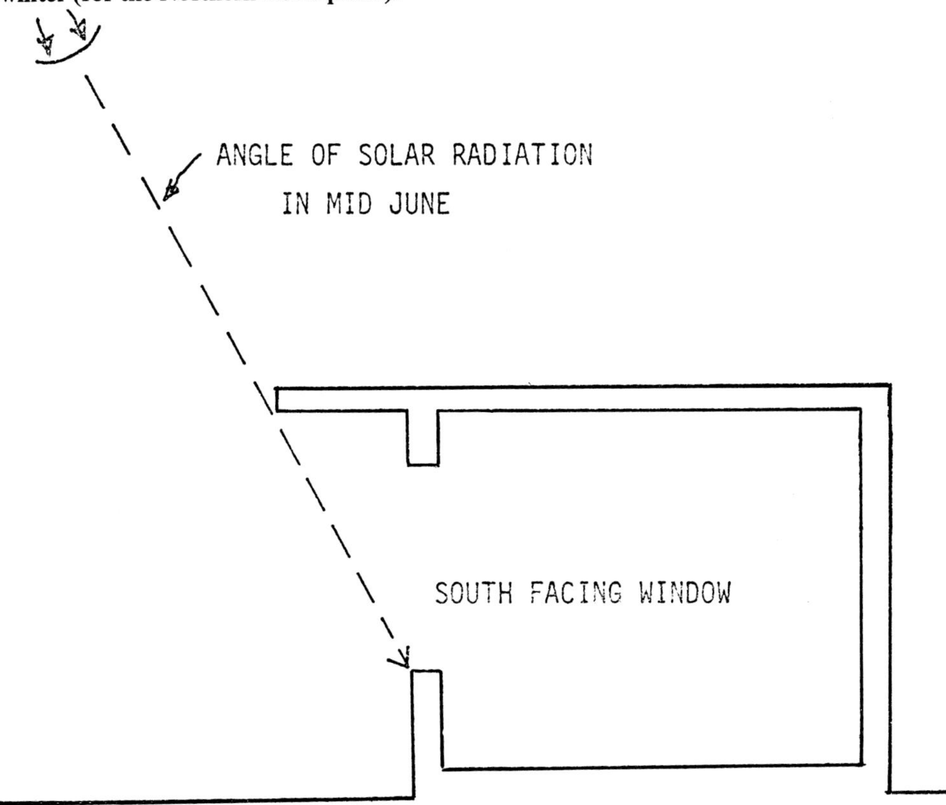

These quotations demonstrate an understanding of the design features which enable buildings to take advantage of warmth from the sun in the winter and to avoid the overheating which might take place during the summer. At the present time we may well be forgiven for wondering whether designers are as well equipped with understanding of the principles involved as they were 2000 years ago. Fig 2 shows the principles used by Xenophon.

Example 2 – Energy conservation in temperate winters

The thatched cottage, apart from its rural charm, makes extremely economical use of local materials to achieve energy conservation. Most of the surface exposed to the weather is of highly insulating thatch. The walls, although not of insulating material are thick, thereby providing considerable thermal resistance and they are of limited area. Windows are small. Ceiling heights are low. The fireplace is centrally located and contributes heat to the rooms on the upper floor. A single fire, therefore, is able to maintain comfort throughout the dwelling. Fig 3 shows a typical thatched cottage.

Example 3 – Summer comfort in a hot dry climate

The courtyard houses typical of North Africa demonstrate the use of several different principles to achieve comfort in a typical hot dry type of climate where the sun shines during the day but the temperature falls at night because of radiation to the clear sky. Two storeys of rooms are grouped around a courtyard. Openings to the exterior are reduced to a minimum on the ground floor. During the night, radiation to the clear sky cools the air in the courtyard which flows through into the ground floor rooms. During the day the sun shines but at first it cannot penetrate into the whole of the court and, even when it does so, the pool of cool air is not immediately dissipated since much of it is protected from the sun by the ground floor rooms.

Figure 3
A thatched cottage.

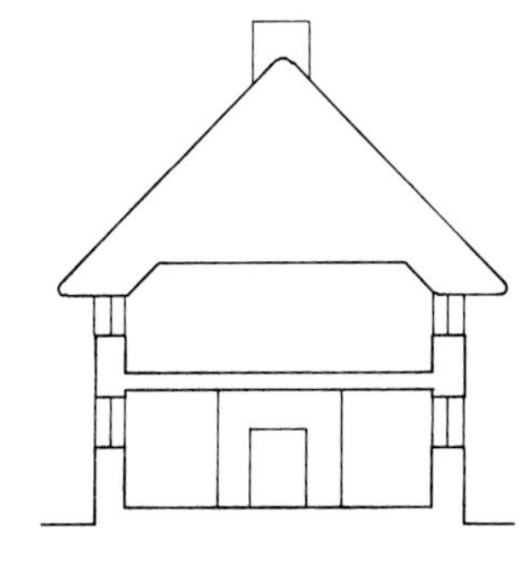

The walls of these houses are typically white painted since this reduces, to some degree, the absorption of solar radiation. They are also of heavy construction so that the solar heat is absorbed into the wall, taking a number of hours to penetrate from the outside to the interior. The performance of the wall will be ideal if it allows the sun's heat to reach the inner face during the coolest part of the night. The courtyard is often provided with a pool of water and a fountain. In a dry environment evaporation is rapid and the water in

Figure 4 **A courtyard house from a hot dry climatic area.**

(a) Conditions at night where radiation to the cold night sky is cooling the fabric and creating a pool of cool air in the courtyard and ground floor rooms.

(b) Conditions during the day when the heat of the sun is being reflected by white surfaces and absorbed by the heavy fabric of the building. The cool air in the ground floor rooms still remains, shade is provided in the courtyard which also has evaporative cooling.

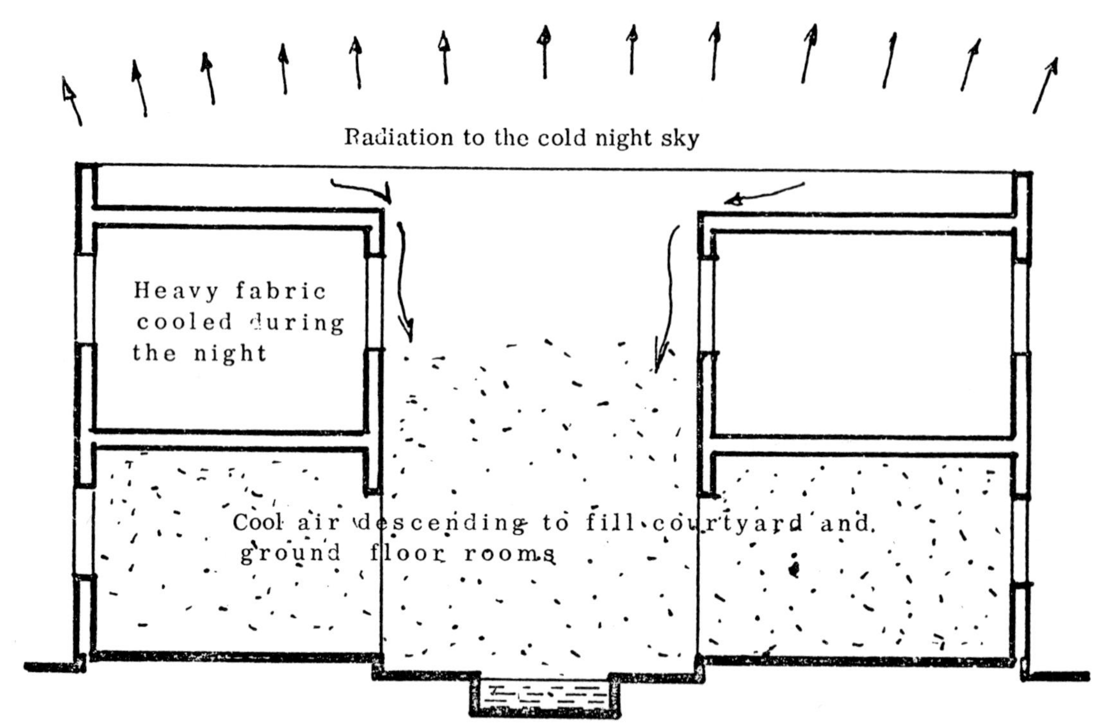

CONDITIONS AT NIGHT

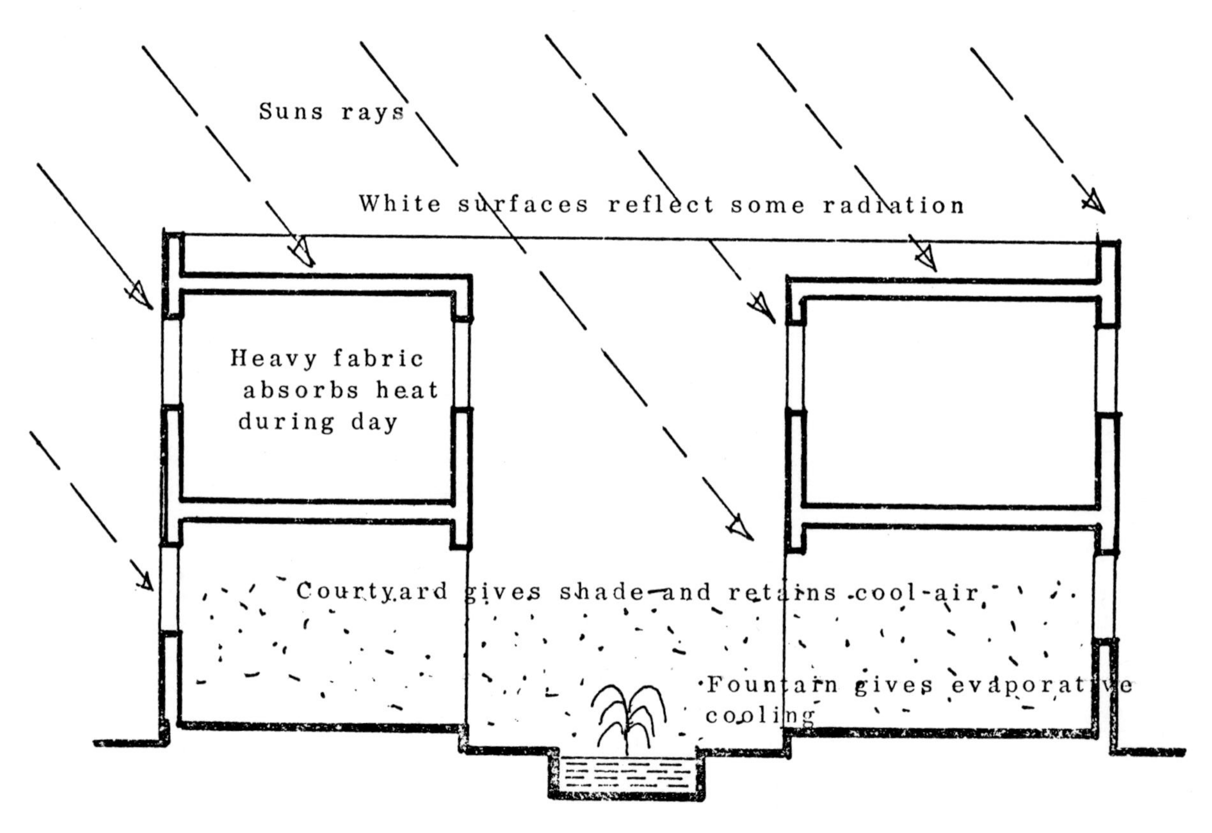

CONDITIONS DURING DAY

evaporating takes up significant quantities of heat thereby reducing the temperature of the air. These factors acting together achieve a remarkable level of comfort in a climate with harsh levels of ambient energy. Fig 4a and 4b show the night time and day time conditions.

Example 4 – Comfort in a hot humid climate

The options open to the traditional designer in a hot humid climate are extremely limited. The only available way to ameliorate thermal discomfort is to increase the rate of air flow over the skin and thereby to increase the evaporation rate of sweat. The veranda type dwelling set on stilts achieves this remarkably well. Much of the day to day living goes on on the veranda, away from the prevailing angle of the sun and takes place in outdoor levels of air movement. Circulation is along the verandas allowing the enclosed volume to be only one room deep. The walls are louvred and a substantial amount of air movement can take place through the enclosed areas. Air temperatures in this type of climate drop only slightly at night. The lightweight materials of construction themselves cool rapidly and allow the benefits of the slightly reduced temperatures to be enjoyed inside whereas more massive construction would retain sufficient heat from the day to keep the air warm during the night. Fig 5 shows the building solution to the hot humid climate.

Modern technological possibilities

The resources available to modern designers are enormously greater than those of the past. Walls, which in this country were of necessity of masonry construction with relatively small openings can now be formed in a range of materials from very dense concrete to lightweight cladding, or even complete glazing. Electric fans and fluorescent lamps mean that buildings need not be constrained in their form by requirements for natural light and ventilation. Many forms of thermal installations exist with widely different characteristics. It is possible to provide cooling as well as heating.

It might be thought that these resources would have been organised to achieve very much more effective balances with external energy than existed in the past. So far this has not occurred. This is, in great part, due to the recent abundance of cheap energy which not only allows thermal installations to be operated but also is the explanation of their developments.

The conservatism of the building industry and particularly of the building design professions is also an important influence. The professions became established in the nineteenth century when, in the great majority of buildings, no decisions were required about the thermal performance of the building. Architects did give some thought to orientation of rooms in the days before universal central heating rendered this unnecessary, but apart from this the limited range of choice of the materials and construction and the sizes of openings meant that designers did not have to consider the thermal consequences of these decisions since these consequences were an inevitable result of the limited range of constructional possibilities.

A pattern of professions developed which did not include any member trained to undertake the fundamental thermal design of the basic form, fabric and fenestration of buildings. Heating engineers were able to evaluate the losses or gains to be anticipated in designs already prepared, but this was in order to size the plant and was not a skill to be deployed in the early fundamental stages of design when the crucial decisions governing

Figure 5

A house from a hot humid climate where increased air movement is the only means of improving comfort and where a lightweight construction takes advantage of the small temperature variation between day and night.

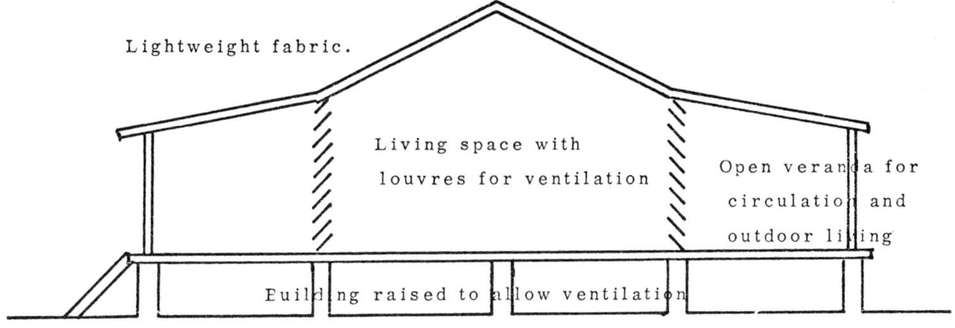

thermal performance are made, long before conventional engineering analysis can be applied. There was no major research into the design of buildings for thermal efficiency. Although there are notable exceptions the basic situation remains the same today.

Traditionally developments in design have proceeded by trial and error over very long periods of time. Today this system is no longer effective because the rate of technical change is such that both the problems and the technology available are likely to have changed before it is possible to design, construct and observe a building for long enough to obtain results which could influence future design. To design thermally in the present context of rapid change it is therefore necessary to be able to predict the thermal performance of proposed buildings. The predictions must cover summer overheating as well as winter comfort. They must also include the energy consumption over the heating season. This involves a degree of accuracy in prediction of much greater precision than that required for sizing plant.

As yet the available prediction techniques are limited and designers not accustomed to evaluate these factors regard them with suspicion. The potentialities, for building design are, therefore, largely unrealised. Indeed in many cases use of new forms or materials without evaluation of the energy consequences has led to unexpected and unfortunate results.

These technological problems could be solved relatively easily by research and development studies. The savings, in relation to the outlay involved, would be extremely large. The scale of effort, however, remains small partly because of the inertia in the industry and partly because of the orientation in this country towards the funding of research in terms of classical scientific objectives. New concepts of the form and functioning of buildings are not regarded as significant when tested by these criteria. The idea that fundamental scientific investigations should be supported by the community while development work which can lead to commercial exploitation can be financed by industry has much to commend it. Unfortunately although buildings are produced and, in effect, sold, good standards of building design are not, in the present organisation of society, a marketable commodity which commercial organisations are likely to feel economically justified in pursuing. It is as necessary for the community to support this type of work as it is for scientific studies. Research should be on a much greater scale than present if it is desired to improve the standards of design of building which, in recent years, has been the subject of such heavy and often well justified criticism in many other aspects as well as the environmental ones of present concern.

COMFORT AND BUILDING DESIGN CRITERIA

The considerations described might lead one to the conclusion that criteria for individual physiological comfort would be well explored while the criteria to assess the performance of buildings themselves would be less developed and this is undoubtedly the case.

Many years of research have determined appropriate ambient energy levels to ensure comfort for the occupants of buildings and the effective performance of many types of task. Many of these concepts are very complex and unite a range of relevant factors into an integrated physiological concept. The principle is well demonstrated by the work done in the field of thermal comfort. Starting with 'wind chill' in the early nineteenth century, which recognised the combined effects of air temperature and air movement in controlling body heat loss, a series of indices have been developed which in whole or part relate the effects of the following to comfort levels:

air temperature
air movement
mean radiant temperature
relative humidity
level of activity
amount of clothing

The indices include Kata cooling power, globe thermometer temperature, effective temperature, equivalent temperature, equivalent warmth, resultant temperature, equatorial comfort index, and predicted mean vote.

A series of special instruments have been developed to measure environments in terms of the indices tested. They include:

Heberden thermometer
Kata thermometer
Vernon globe thermometer
Dufton eupatheoscope
Comfort meter

A detailed discussion of thermal comfort is not appropriate in the present context. It is apparent, however, that a very great deal of scientific effort has been devoted to this subject over a long period of time, and similar, if less extensive, work has gone on in the other energy fields.

The amount of attention given to the thermal performance of buildings is very much smaller. Approximate techniques for estimating plant sizes have existed for some time but it is only in very recent years that any method has been available to predict the variations of temperature of the interior of the building as a result of interaction of external influences and the thermal properties of the fabric of the building. Apart from the obvious, but not very practical ideas of spherical or cubical buildings there are no accepted overall concepts of building form and fabric which provide a proper balance of the various elements of building design to give optimum design balances between economy and performance. While it is possible to identify the factors which govern the thermal performance of buildings and to a limited degree to evaluate them there are no criteria which enable us to decide whether a building is carrying out its thermal tasks in an efficient and economical way.

Table 1 sets out the overall scope of the problem. It lists the main aspects of ambient energy and tabulates their direct effects on buildings and occupants in relation to the building design factors which give rise to the behaviour of the building itself and govern the degree to which the comfort and efficiency of occupants can be maintained and the economy of materials and of the energy required for environmental control installations.

SIGNIFICANCE OF BUILDING DESIGN FACTORS IN ENERGY UTILISATION

While the criteria for human comfort and, to a lesser degree, human performance are dictated by ideal rather than economic criteria it is inherent in any criterion for the design of buildings that performance is achieved with economy of resource. At present we are acutely conscious of fuels as a particular resource to be used efficiently and to be conserved.

It is desirable to consider the relative importance of the various building design factors which determine the performance of buildings in relation to ambient energy. It is not possible to lay down universal rules since the importance of the various factors is not fixed but will vary with different designs and functions of buildings. It is possible, however, to indicate the general order of importance and the conditions which determine this.

There are several modes required for the design of buildings in relation to ambient energy. First is the positive mode needed when it is desired to capture and employ usefully energy available in the external environment. Second is the negative one where the occupants must be protected against excessive levels of external energy. Third is the conservation mode where the need is to minimise the consumption of energy and its loss to the external environment.

The positive mode is exemplified by buildings designed to take advantage of solar heat in the winter and to promote air movement for improvement of comfort. The negative

Table 1 **Aspects of ambient energy, their effects on buildings and occupants and the main building design factors which govern the effects on occupants.**

Aspects of Ambient Energy	*Direct Effects on Buildings (a)*	*Direct Effects on Occupants (b)*	*Building Design Factors Governing Effects on Occupants (c)*
Solar heat Air temp	Thermal movement Deterioration of finishes	Thermal comfort Air temp MRT (Air movement) (RH)	Surface area Insulation Capacity Volume Height Fenestration Orientation
Air movement	Pressure differential across windows and ventilation intakes and exhausts Structural loading Noise generation	Ventilation rate Comfort velocity Removal of flue effluents Thermal comfort	Crackage Infiltration Volume Height Windows
Moisture	Moisture movement Deterioration of materials Reduction of insulation Condensation (in conjunction with heat and air movement)	Thermal comfort	
Sunlight	Deterioration of materials	Visual quality Psychological satisfaction	Block spacing Floor to ceiling height Block width Fenestration Orientation
Noise		Annoyance Hearing difficulties	Planning to separate noisy and quiet activities Attenuation by materials Fenestration

mode can be exemplified by design to minimise heat gain or to protect occupants from excessive air movement. The difficulty of design is well demonstrated by the fact that the positive and negative aspects are in conflict. A design which achieves excellent air movement for summer comfort may present comfort and economic difficulties in winter. It is difficult to exemplify the conservation mode of design. A number of buildings having excellent features in this respect have been designed but it is too early to draw general conclusions about conservation performance.

Positive design for ambient energy

There is only one aspect of the natural environment which can contribute energy in the positive sense to buildings without the agency of an installation. This is the direct radiation from the sun. Windows allow this radiation to pass into the building. During the summer, of course, this may give rise to excessive temperatures and adequate ventilation and in some cases screening may be needed to prevent excessive temperatures. In winter, however, the rays of the sun penetrating to the interior of the building through the window can reduce the rate of input required from the heating installation, making significant contribution to the total energy requirement for heating.

Figure 6 **The behaviour of solar radiation impinging on window glass.**

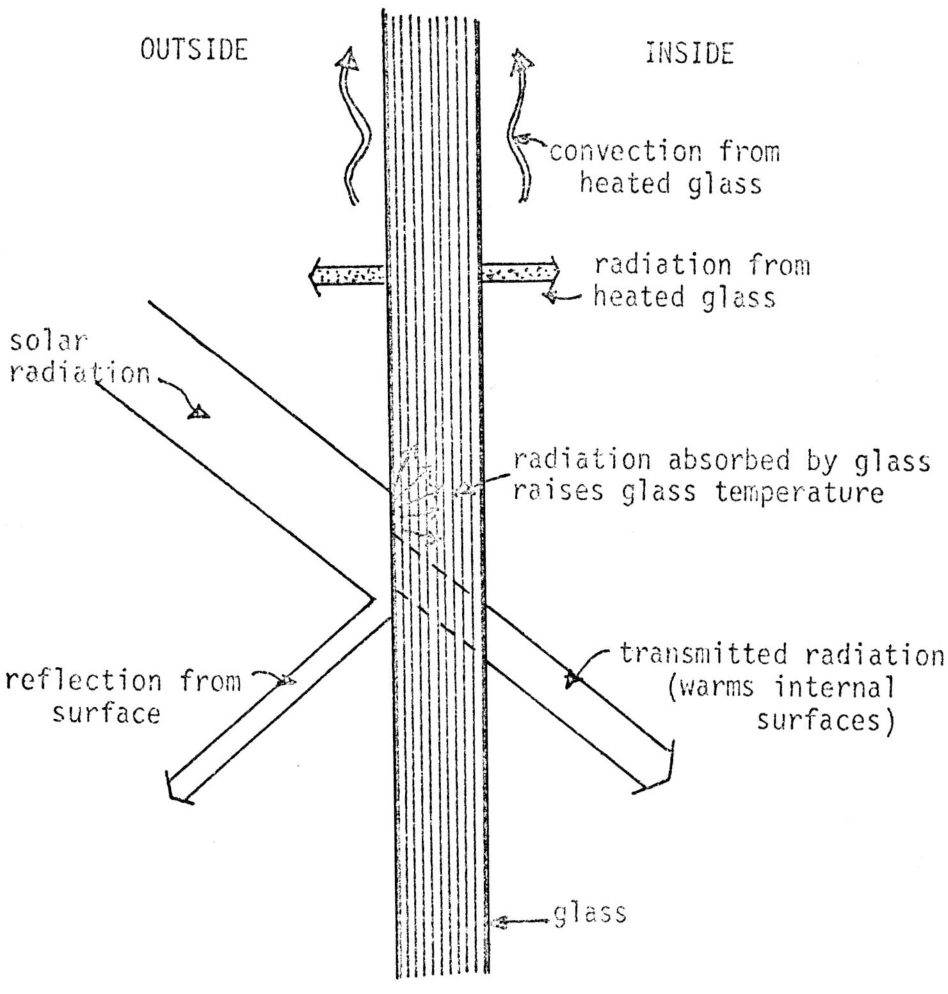

Figure 7 **Solar radiation in rooms.**

Massive floor construction will absorb heat rapidly. Lightweight or carpeted floors will reach higher surface temperatures and a larger proportion of the heat will be immediately liberated into the room giving rise to higher peak temperatures.

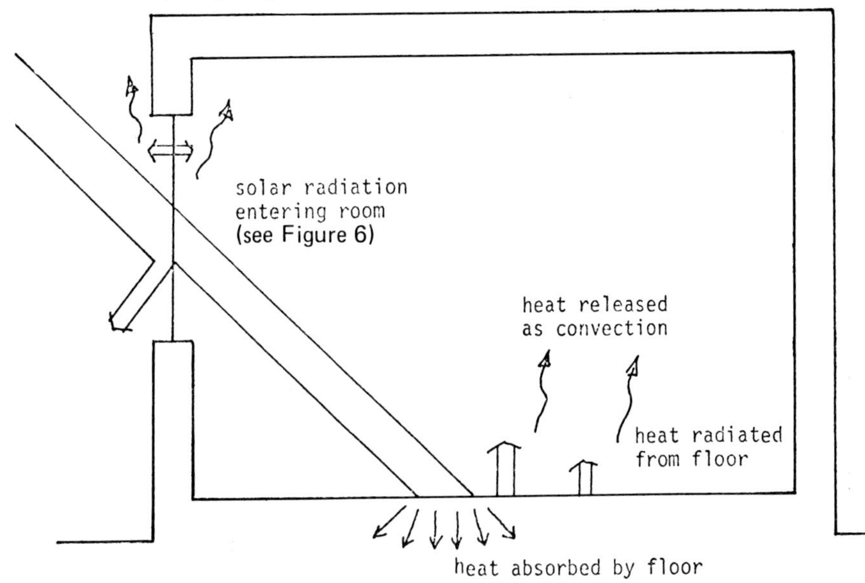

Figure 8a Variation of external and internal air temperature together with monthly losses sustained through one square metre of uncurtained single glazing and one square metre of single glazing curtained at night.

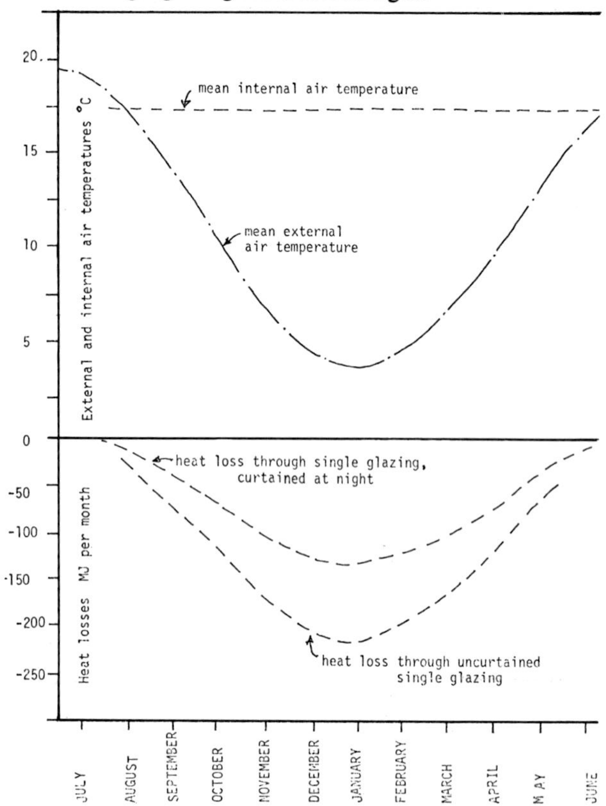

Figure 8b Maximum possible and actual solar gains through one square metre of south facing clear glass.

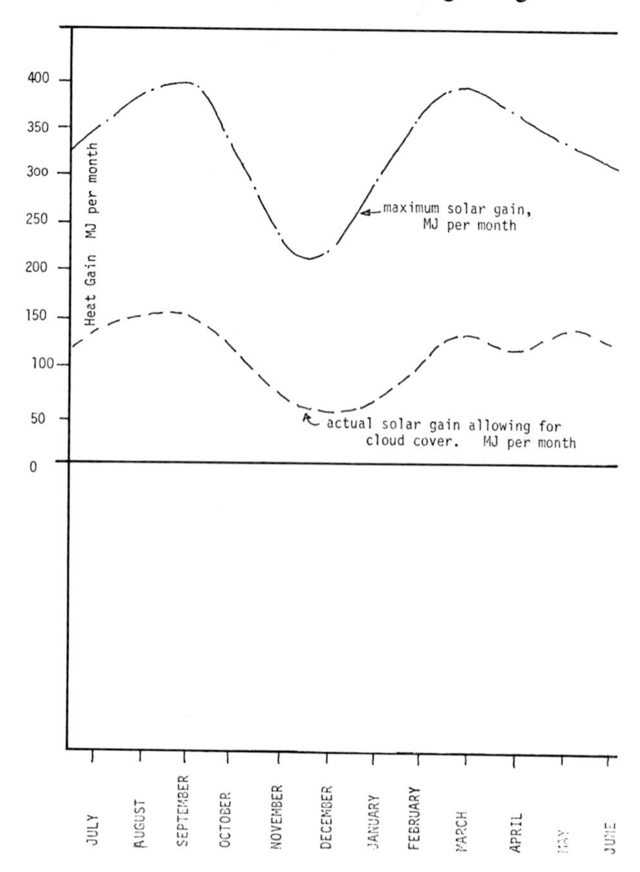

Figure 8c Net heat balance through one square metre of unobstructed south facing single glazing. The mean internal air temperature is 17.1°C and the windows are curtained at night.

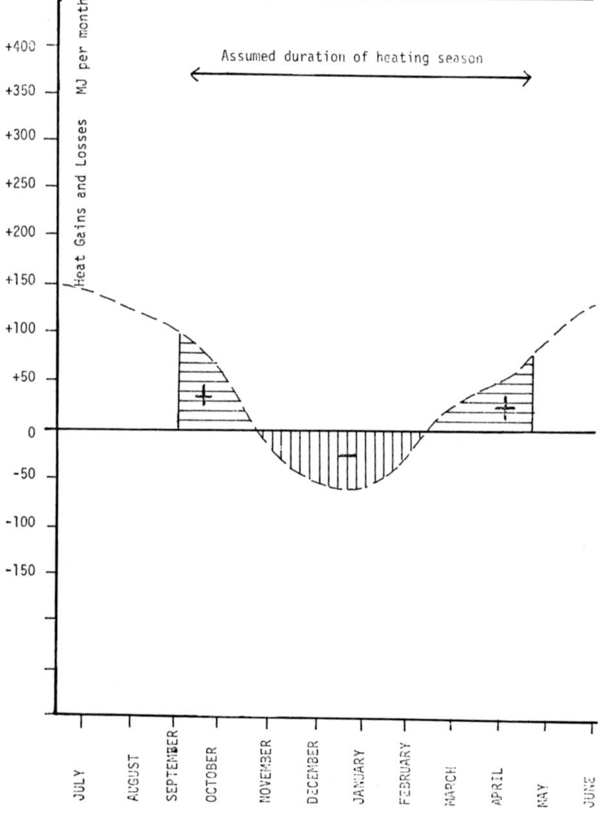

Figure 9 Variation of net heat balance with orientation for one square metre unobstructed clear glass window, curtained at night with a 17.5°C mean internal temperature. *(Originally published in "The Architects' Journal")*

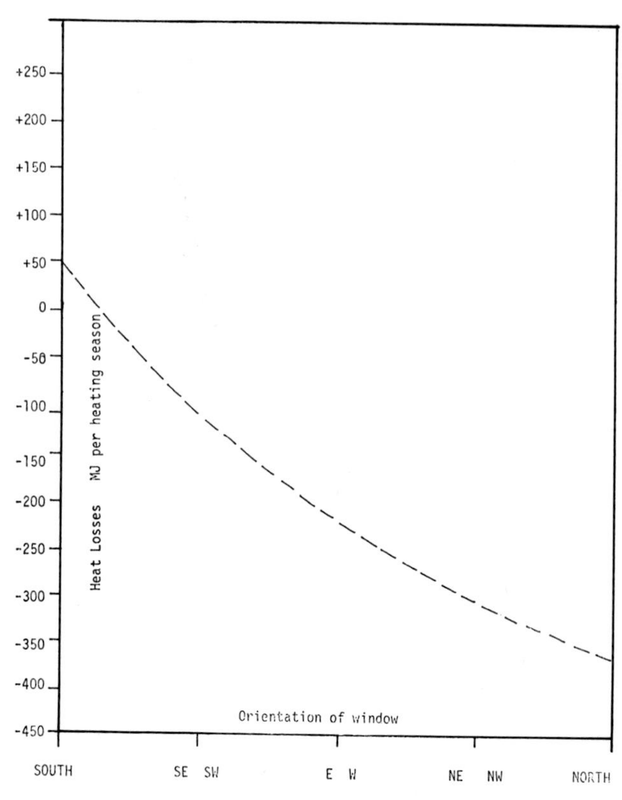

Table 2 **Variation with orientation of the net heat loss through windows. Heat balance for 1 m^2 single glazing for heating season.**

Orientation	*Heat Gain MJ (BRE)*	*Nett Heat Loss MJ*
South	680	0
East and West	410	270
North	250	430

Note: Heat loss through 1 m^2 single glazing = $3.4 \times 10 \times 2 \times 10^6$ J
= 680 MJ

Where:
U = 3.4 $W/m^2 {}^\circ C$ (BRE, curtained window)
Average Temp Diff for heating season for centrally heated house = $10^\circ C$ (BRE)
Length of heating season = 2×10^6 seconds.

The total effect of this can be extremely significant. Fig 6 shows the effects of solar radiation passing through glass. As can be seen the great majority of the energy passes through the glass and reaches the interior. Fig 7 shows the effects produced by the solar radiation in the interior of a room. Clearly the success of utilising the solar radiation depends on the size and orientation of the windows and also the materials of construction and finishes in the room. Fig 8 shows, how, for a south facing window, curtained at night in a domestic situation, with single glazing the losses through the window are balanced by the gains.

Table 2 shows the variation of gains with orientation. It will be observed that it is only with southern exposure that there is likely to be positive gain during the winter. Other orientations will give a net loss of heat over the heating season but there is still a major difference in the amount of loss depending on the orientation. Fig 9 shows how the net heat balance for a window varies with orientation.

The conclusion which designers should draw from these aspects of heat gain is not that one should immediately make windows on southern elevations larger. It is, however, clear that efforts should be made when planning the building to ensure that windows are located with the best possible orientation to minimise heat loss. It is very apparent at present when studying domestic buildings and particularly in new towns where large numbers of houses are being built together, that the orientation is completely at random. One can observe standard designs repeated many times which have large windows in one particular facade. It would be natural to expect that their orientation would be related to sunshine, or to view. In many cases this does not appear to be the case. Large windows face north, south, east or west with no apparent purpose other than sculptural effect. Central heating has made many designers totally unconscious of the importance of orientation.

The materials and finishes in the interior of a building are very important if solar energy is to be utilised to its full. If lightweight finishes are used the rays of the sun will impinge upon them, warm them rapidly and give a very rapid increase in internal air temperature. Even in the winter this may give rise to overheating and the excess energy will have to be dissipated by opening windows. If more massive materials, with higher thermal capacity, are used the heat from the solar radiation will be absorbed to a considerable degree into the material. This not only reduces the immediate rise in temperature but also allows the excess heat to be liberated steadily later in the day when the direct rays of the sun have ceased.

St George's School Extension, at Wallasey, which took advantage of these principles, was designed before it was possible to evaluate their effects. Fig 10 shows a section of the construction which is massive with thick external insulation. Ventilation is carefully controlled and large windows face south across unobstructed ground. Heat gains are drawn from the sun, the occupants and the tungsten lighting. For several years the building gave acceptable thermal performance without the use of any heating installation other than the

Figure 10

A section through the extension to St George's School, Wallasey, showing the massive construction, thick external insulation and double window wall.

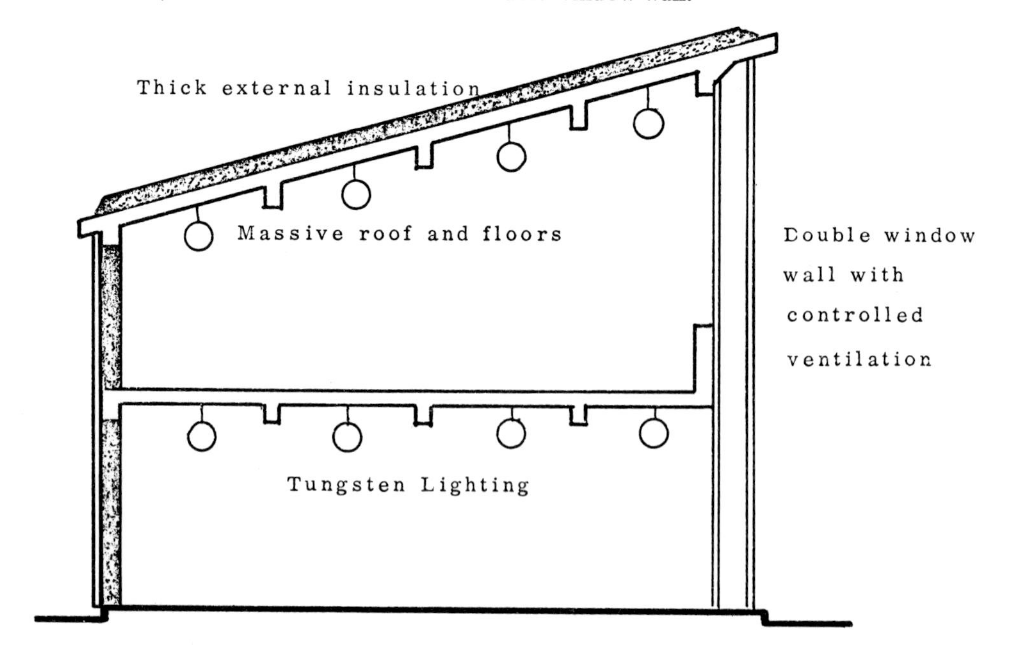

Figure 11

Histograms showing the peak environmental temperatures achieved in heavyweight and lightweight rooms with various measures for solar control.

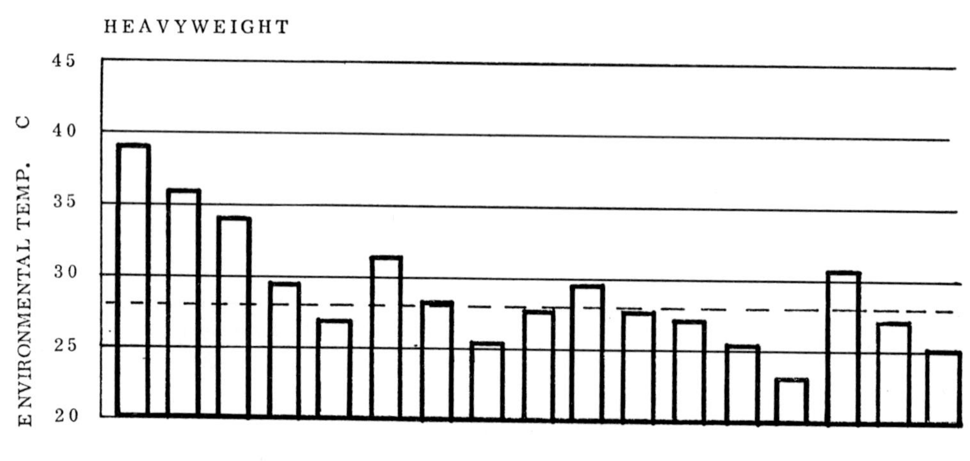

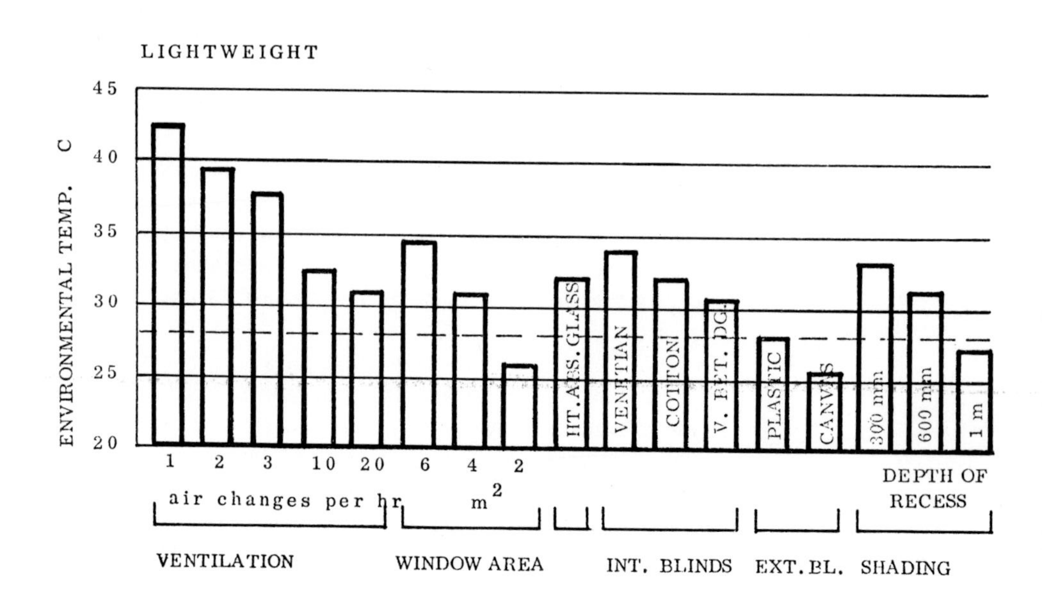

lighting. Although the school represents a specimen case and it is not practicable to rely solely on solar and wild heat gains in other buildings it nevertheless demonstrates dramatically the importance of the form, fabric and fenestration of the buildings in determining thermal performance.

Some methods now exist which enable prediction of the performance of the building fabric. At the present time the Admittance Method developed by the Building Research Establishment does enable some comparative estimates to be made of the effects of internal materials and finishes on the temperatures likely to be reached. Fig 11 shows two histograms demonstrating the effects on peak internal temperatures of a variety of design variations which may be made to ventilation and window arrangements. One of the histograms shows the condition for a low thermal capacity interior and the other deals with one of high thermal capacity.

It will be observed that in the high thermal capacity interior the temperature, except for very low rates of ventilation, rarely rises above the 28°C taken as the maximum level for summer comfort. In the low thermal capacity interior, however, the temperature is significantly above this level for most of the cases illustrated. The example taken is related to summer overheating but the principle demonstrated is a valid one for winter conditions. The difference in the peak temperatures between the low and high thermal capacity buildings is entirely due to absorption in the materials of construction. And, in winter conditions, particularly at the beginning and end of the heating season, this ability of the construction to absorb solar radiation will contribute both to comfort and to economy of heating.

The total saving of energy which can be achieved by these means will not, in most cases, be very large. On the other hand it is a saving which can be made without any investment in additional plant or equipment or any increase in the building cost. It gives the opportunity of infinite cost benefit.

Negative design: Prevention of extremes

Every aspect of ambient energy in the external environment must be regarded, in the English climate, as presenting an extreme in relation to human comfort. If this were not so, apart from privacy, there would hardly be a need for buildings at all. The mere existence of a shell is effective in dealing with several aspects of the problem. Precipitation and wind can be thus controlled. Solar radiation and air temperature present a more complex problem. To perform effectively the shell of the building must have some insulating properties and also some thermal capacity. The windows necessary to admit light must be shaped and oriented to admit sunshine to make the interior pleasant in winter but not to such a degree that overheating takes place in the summer. At present although they are not used as widely as they should be there are design methods available which enable these factors to be taken into account and effective design results achieved.

Conservation: Design to save energy

Very large amounts of energy are now input into buildings. There are sophisticated and adequate techniques for ensuring that the plant required to input this energy is adequately sized. Very little attention has, however, been given to the problem of designing to minimise energy consumption over the whole heating season. This involves consideration, not only of the heating installation but also of the form of the building, materials of construction, orientation and fenestration.

Two aspects of the external environment have a major influence on energy losses and two others a less important but nevertheless significant influence. The most important factors are external air temperature and wind. The less important factors mean radiant conditions and moisture, either in the form of vapour in the air or precipitation. Low external air temperatures cause heat losses to take place directly through the fabric of the building, they mean that the air necessary for ventilation must be warmed through a greater range of temperature and, by means of the classical thermosyphon phenomenon they give rise to increased rates of ventilation because of 'stack effect' within the building which begins to act as a flue drawing cold air in at the bottom and exhausting hot air at the top.

This phenomenon becomes extremely marked in multi-storey commercial buildings many of which have to be provided with mechanical ventilation in order to overcome this particular problem. Higher rates of air movement increase heat losses from the external surfaces of the building and also contribute to high rates of ventilation. Radiation to the cold night's sky can produce significant reductions in temperature on the outer surfaces of buildings. Clear skies often occur during cold spells but this phenomenon is not likely to have a major effect on energy consumption itself. It may well, however, contribute to the incidence of condensation which can be a major problem in buildings. Moisture in the atmosphere can contribute to heat losses by increasing the conductivity of building materials.

There are clearly two ways in which these problems can be minimised. The first is the siting of the building and secondly the design of the building itself and its form, fabric and fenestration. Although nineteenth century text books of building design discuss the problems of siting as one of the responsibilities of the architect it would be rare today for the designer of the building to have very much influence on its siting. In ages when the pace of technological change was more gentle and travel governed by the limits of human and animal energy designers were familiar with the areas in which they built and were able to site buildings advantageously.

The situation is now very different and in selecting a site few people give detailed consideration to the microclimatic features. Even if there were a desire to take these matters into account it is rare that designers would be sufficiently familiar with the area to work from experience. Meteorological data, which is not in any case connected with the needs of building in mind, is not sufficiently detailed to provide adequate guide. It is impossible to say what percentage of energy might be saved by taking account of microclimatic conditions. It would not normally be reasonable to indulge in major expense such as additional roads and services and the consequent energy consequences in order to achieve what may be a relatively small energy saving in the buildings erected. However, there can be no doubt that careful attention to these factors by town planners and designers would achieve a measurable, if small saving. Even a fraction of a percent, although hardly significant in the context of an individual building, if taken over the building stock as a whole represents a very substantial amount of energy.

Since external energy losses are governed by the area of the external skin and the volume to be ventilated it follows that the shape of the building which will govern the area of external skin to useful interior space and will, to a considerable degree control the influences of stack effect and wind on ventilation is a major factor in designing in relation to external conditions. Within certain limits shape does not have a major influence on heat losses. Fig 12 shows that for a two storey building variations in the plan form from a square shape to one with a ratio of 3:1 between length and width, even when additional offsets are included, does not have sufficient effect upon heat losses to justify any major departure from what would otherwise be an appropriate solution. It does not seem likely that a building which could have an acceptable functional disposition within a square plan shape could easily be accommodated in a shape with an aspect ratio of 1:4 or more.

Such a re-arrangement would very likely produce a substantial increase in the circulation space necessary and thus increase the size to accommodate the same function. When more extreme variations of shape are possible, such as might be the case when multi-storey construction is a possible alternative, the energy implications are very much more serious. It is not meaningful in these cases to make single geometrical comparisons, although it is clear that the external skin of point or slab multi-storey blocks will usually be greater for the same floor area. Other factors begin to enter the equation all of which result in greater losses of energy to the external environment.

The multi-storey building suffers the disadvantages of increased exposure. It suffers

Figure 12 **The variation of heat required in relation to shape and volume.** *(Originally published in "The Architects' Journal")*

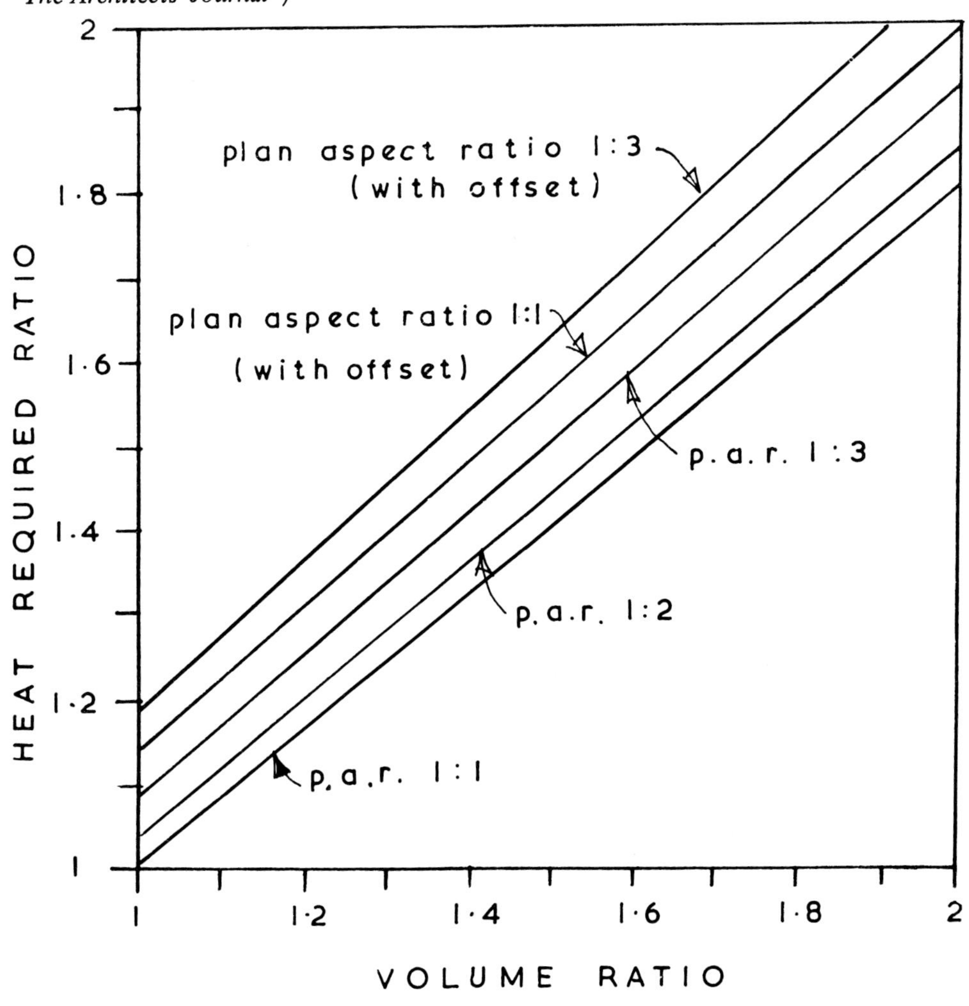

Figure 13 **Output from hybrid computer showing cumulative energy consumption.**

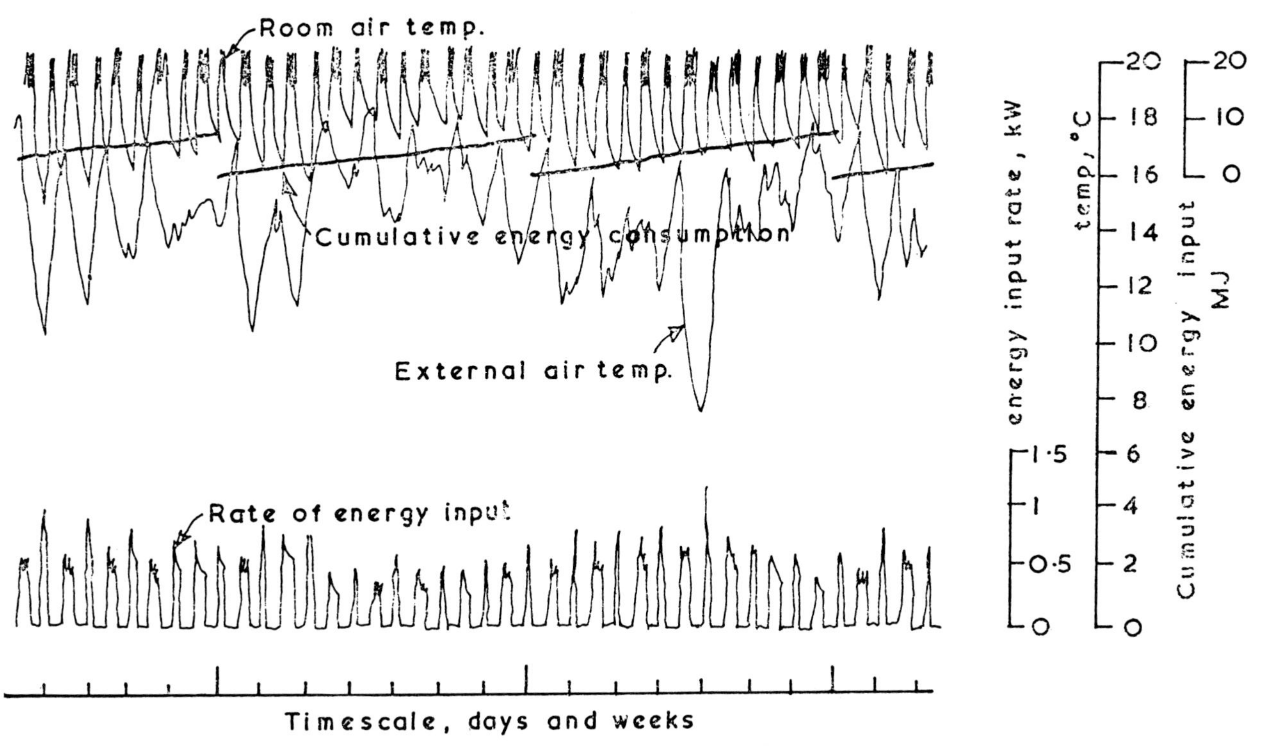

acutely from stack effects and it is difficult to control ventilation effectively. The ventilation problems and the results of unprotected exposure to solar radiation very often result in a requirement for air conditioning which has very distinct energy implications. In this type of building also the area devoted to circulation may be substantially increased, not only by the needs of circulation itself but also by the requirements for fire fighting and escape. This, together with the increased accommodation for services, is likely to result in an increase in the overall building size to accommodate the same functional area.

Fig 12 does demonstrate, however, that consumption of energy is directly related to the volume of the building. Thus designers who plan and utilise space efficiently and economically benefit, not only in capital cost, but also in energy conservation. It is particularly effective since it reduces both fabric and ventilation losses and presents a situation where the wild heat gains represent a greater proportion of the total energy requirement of the building.

Ventilation is influenced by form and exposure but can also be controlled by attention to the sealing of the fabric and tightness of window closing. In many existing buildings very significant energy savings could be made by sealing the interior from the exterior wind influences. Care must, however, be taken when considering reducing ventilation rates to conserve energy. Condensation might begin to take place. The moisture regime should be investigated very carefully in any building where low ventilation rates may be used. Summer overheating is also likely unless it is possible to increase the rate of ventilation very greatly in the summer. Table 3 shows some of the methods of reducing energy losses and the mechanisms of loss which they influence. It is clear that methods which control both fabric and ventilation losses are, if they can be used, more effective than those which only influence one of these mechanisms.

Table 3 **Effectiveness of Energy Conservation Methods.**

	Method of Reducing Energy Loss				
Mechanisms of Loss	*Reduce Air Temp*	*Reduce volume*	*Improve shape*	*Increase insulation*	*Reduce ventilation*
Fabric	X	X	X	X	
Ventilation	X	X			X

ACTION FOR DESIGNERS, SPONSORS AND CRITICS

The traditional building solution to control ambient energy developed by trial and error over a long period of time. Modern designers do not have this opportunity. Solutions must be found to new problems. Because of the recent availability of cheap fuel little interest has been taken in the thermal performance of buildings for the last quarter century with the inevitable consequences upon knowledge, skill and experience both in research and design. It is not practicable to wait for feedback from experimental buildings.

At present there are few methods available for predicting the thermal behaviour of buildings with sufficient precision to be useful in giving a basis for improving the utilisation of ambient energy. Fig 13 shows the degree of precision and detail which can be achieved by the use of hybrid computing techniques. The chart shows how variations in internal temperature can be recorded together with the cumulative energy requirement, taking into account the variations of internal and external temperature, the characteristics of the heating installation and its controls and the thermal capacity as well as the thermal resistance of the building construction.

Until such studies have been carried to the point where the results are available in the form of convenient guides for designers ideal solutions will be difficult to achieve or assess. All of the worst cases of excessive energy consumption and unpleasant comfort conditions could, however, be eliminated if designers became familiar with the very simple thermal phenomena involved in buildings and gave weight to them during their designs. Unfortunately this does not happen at present in most cases. While one could not anticipate achieving optimum results in this way there is no doubt that there would be a dramatic improvement in general design standards and that all the major thermal failures would be eliminated.

This deals only with thermal comfort and energy conservation. Positive design to incorporate all the ambient energy factors into one overall optimum design solution presents enormous difficulty at the present time. Table 4, however, demonstrates that the energy variables are more easily controlled in most cases when the relevant building design elements are kept to a minimum size. It would be impossible to attempt to design specifically for all the different factors individually.

Table 4

Effects of Design Variables on Internal Climate Conditions

Design Factors	*Result of reduced dimension or quantity in terms of control of internal climate*					
	Solar/ Gain Heat	*Daylight*	*Noise Penetration*	*Natural Ventilation*	*Heat Loss*	
					Fabric	*Vent*
Volume	0	/	0	/	/	/
External Surface Area	/	0	0	0	/	0
Window Area	X	/	/	0	/	/
Floor to Ceiling Height	0	X	/	X	/	/
Depth of Building from Window Wall	X	/	0	/	0	0

Key:
/ signifies improved performance
0 signifies marginal variation of performance
X signifies deterioration in performance

Heroic design solutions such as spherical buildings or beehives without external light or ventilation are fortunately not with us yet and, since they make no positive use of ambient energy may never come into use. It is, however, very encouraging to know that a design which achieves its results with the minimum use of space and building materials in conventional form is likely to lend itself to a useful degree of energy conservation and also to be adaptable to the positive use of ambient energy.

DISCUSSION

J. Harrington-Lynn (Building Regulations Division, D. O. E.)

To make use of solar gains, the heating system and its controls will have to be designed to sense and adjust to the gains. This may only be the cost of extra thermostats, but this cost has to be allowed for. The extra cost in designing to overcome the problem of summer overheating, if glazing is unrestricted in area, will also have to be allowed for. Therefore the simple statement that south facing glazing may have an infinite cost benefit, like most other generalisations cannot be taken at its face value and needs careful qualification.

Professor Burberry's analogue results show the effect of placing the insulation in different positions. For short heating periods, or where more than one heating period occurs during the 24 hours, I would agree that it is extremely difficult to assess these effects, unless analogue or digital computer simulation is used. However good approximations to the analogue results can be obtained, for heating periods of 8-16 hours using the admittance procedure. A recent paper by Billington* demonstrates this approach. For short heating periods, the admittance procedure can be misleading unless adjustments are made to allow for the higher frequencies involved.

*Billington, N. S. 'Thermal Insulation and Thermal Capacity of Buildings'. Proceedings of conference, 'CLIMA 2000' Milan, March 1975.

Professor P. Burberry

Mr Harrington-Lynn is making a very valid point. In giving the energy variation for the change of orientation of 1 m^2 of window I did allow 50% wastage to anticipate this particular point. However, it is very desirable to amplify this. The degree to which solar energy can be utilised depends very considerably upon the nature of the building fabric and the response characteristics of the heating installation. It is desirable that the fabric of the building should be able to absorb the heat of the sun during the period when it is shining into the window, thereby reducing the peak temperature and making the heat available by retransmission into the room after the sun has gone down. It is important that the heating installation can cut off its input to the room in question when the sun is shining. If this does not happen it is likely that the room will overheat, that windows will be opened to reduce the temperature and that not only the sun's heat will be wasted but heat from the installation as well.

I am glad that Mr Harrington-Lynn mentioned the admittance method. This is an extremely valuable technique and if there is anyone concerned with building design who does not use the method he should take steps to remedy this as quickly as possible. It also gives me the opportunity to acknowledge the Building Research Establishment's pioneering work in the analogue prediction of thermal performance of buildings. BRE were the pioneers in this field and one of the outcomes made possible by analogue application was the admittance method.

J. M. Taylor (Atkins Research & Development)

A computer program is commercially available which undertakes the calculation of heat flows and heat losses and gains through different types of building fabric and internally across partitions. Using this program it is possible to optimise the building orientation, the fenestration, the fabric itself and the shape of the building. It can also optimise the plant size and the period of operation. I believe this program is now being further developed to allow evaluation of the position of shadows cast by the sun at hourly intervals through the day. It will therefore be possible to see where the sun actually shines and the effect of surrounding buildings.

Professor P. Burberry

I am not familiar with the program and cannot comment on it specifically. It would be very interesting to run the same series of determinations which yielded the information shown on the chart of comparative insulation thicknesses and positions on the digital program and to discover what the comparison is, In the past comparisons of this kind have always demonstrated that digital computation either required substantial simplifications or took a very long time. It would be very interesting to see whether this situation has changed.

P. Brennan (Greater London Council)

Professor Burberry showed us the importance of the disposition of insulation in a wall to minimise the energy required to maintain comfort within a house, but I wonder if the beneficial effects of allowing a little of the energy to pass through the wall is being overlooked. Perhaps Professor Burberry could comment on the likelihood of increased costs for maintenance of the exterior of a building which has no heat going out through the walls? This question is raised since some function of occupancy tends to preserve building facades, the evidence being that most buildings deteriorate rapidly when they are not occupied. My assumption is that the deterioration is not all caused by small boys throwing stones through the windows, but that the lack of heat is an important factor.

Professor P. Burberry

The problem of positioning insulation has been recognised for some time. If insulation is positioned on the outer face of a wall the weatherproof covering can reach very high temperatures as a result of the sun's ray falling on it and considerable movement and deterioration can result. On the other hand this arrangement results in the thermal capacity of the walls being available internally which is advantageous in the summer and has some advantages in winter when the heating is turned off at night and the heat in the walls can help to maintain some degree of warmth in the interior. In Scandinavia this is recognised in the building regulations and variations in the amount of insulation required are made according to the thermal capacity of the building in order to prevent excessive cooling at night. In terms of economy of heating in the winter, except for buildings with continuous heating, it is most advantageous to have the insulation internally so that the walls do not have to be warmed up again after they have cooled down at night. This results in a lower average temperature for the walls and probably higher moisture contents which can lead to deterioration. The optimum resolution of this problem is clearly a very important one in the thermal design of buildings but as yet it is unsolved. So far as I am aware no serious work is going on to determine the point. This is a major omission.

It is not very satisfactory for a designer to be told that no one knows the answer and I will hazard a suggestion which may represent a useful strategy for present design. I take the view that it is best to design the fabric of the building for winter economy and then to make sure that the ventilation, fenestration and orientation are appropriate to prevent summer overheating. If necessary some control of sun penetration is possible for conditions prevailing in summer by means of fairly simple projections over windows. In this way one may have the advantages of both worlds.

R. Cullen (Architects Design Group, Derby)

I am worried by any suggestion that standards should be reduced. We should use every technique, solar, wind, heat pump, etc at our disposal for improving standards. All these techniques have some potential and at this stage should be positively encouraged rather than discouraged. I am sure that the Arab living in the theoretical Architect dream house illustrated by Peter Burberry is too hot and am equally sure that living in an English country house, even with large double glazed windows facing south, is uncomfortable in the winter. We have not looked sufficiently far ahead, or been sufficiently imaginative.

Professor P. Burberry

It is a very happy thought that we can both economise on energy and be just as comfortable as we were before. I am afraid that my view is a more pessimistic one. Energy is clearly getting more and more expensive and I do not think we can anticipate that we shall be able to economise sufficiently in this way.

The past 25 years have been very unusual in that they have been a period of cheap energy. The normal situation in the past was that energy was expensive and indeed for many centuries was in extremely short supply. I think that it is inevitable for the future that we shall be making new balances between the amount of space we are prepared to have, the amount of clothes we wear and the amount we are prepared to spend on energy.

Dr. A. F. C. Sherratt (Thames Polytechnic)

I would take issue with Professor Burberry's suggestion that reduced standards of environmental space and temperature might be needed as a way of reducing our energy demands. The Parker Morris report* in 1961 emphasised that space is only usable in winter if it is heated. The alternative is to congregate in one heated room and use the remainder of the dwelling very little. The *minima* laid down by Parker Morris have since been adopted as the standard provision in local authority dwellings which indicates the undesirability of reducing heating standards and with them standards of living. To return to the single usable room and the long winter wait for the summer when the whole of the dwelling could be used is unthinkable.

Similarly, the environmental standards in offices are as low as is likely to be achievable in the seventies or eighties. At one time conditions were very much worse but to go back to them would be a retrograde step. One way to increase the efficiency of office buildings would be to use them at night as well as in the day time. To make a change of this magnitude would be tremendously complex sociologically and administratively and is unlikely to gain sufficient support for success.

We may grow to require temperatures which are too high. This happened in the USA where temperatures are required which are higher than those acceptable in the UK. Any trend in this direction should certainly be discouraged. The answer to our problems is not to cut the temperature of working spaces and necessarily the standard of environment to which we have become accustomed. The answer is to use our technological expertise to use less energy to achieve the same standard of environment currently enjoyed.

* Ministry of Housing and Local Government "Homes for Today and Tomorrow". Committee report – Chairman Sir Parker Morris HMSO 1961.

Professor P. Burberry

My view is a more dismal one in that I think energy will become more expensive and there will come to be a new balance in the amount of space we are prepared to heat and so on. If you look back, the past 25 years of cheap energy have been a deviation from the norm. By and large, energy has been expensive in the past, desperately expensive very often in the days before coal, when it was necessary to go out and collect wood. It would be excellent if we could make small savings and still have all the amenities that we enjoy at present, but my view is that energy will continue to become more expensive and as it does so, we shall make a new balance. As central heating costs rise, we shall put on more clothes. We will adjust automatically. People will make it in their lives and I think in building design. I am not expecting fusion to emerge at any minute and nuclear power gives me the horrors. I think our fundamental problem is how to reduce the population to a level at which tree growth will sustain us.

R. Cullen

The National Aeronautics and Space Administration, USA, are investigating the use of natural photosynthesis. Water lilies, which are very fast growing, are being used firstly as a means of water purification, secondly to make methane gas, and thirdly compost. By converting plant material to gas, the winter and summer problem is resolved because the gas can be liquefied and stored. This produces a very interesting ecological cycle and this new technology, developed in the space programme, has not been mentioned. It has only been suggested, rather loosely, that we might grow trees again and burn the wood.

Professor P. Burberry

It is important we consider which standards we are prepared to reduce. Marie Antoinette refused to reduce standards but the result of this was a forced reduction. If we insist on pouring energy into our buildings we may find ourselves running out, with very painful consequences. The only way in which we can avoid the logic of this at present is to assume that something is going to turn up. We might describe it as the Mr Micawber syndrome. Fusion energy perhaps demonstrates the dangers of this, 25 years ago it was almost with us, now no one is so optimistic. There is a great deal of very erudite work going on throughout the world into various ways of getting energy from plant growth and refuse as well as the wind, sun and waves but at the moment none of these will solve our problems. We have to decide whether to go marching on with energy consumption at full bore assuming that something will turn up or whether we will moderate our consumption.

K. G. Peters (British Gas)
written comment

In his presentation, Professor Burberry introduced a table giving relative heat flows through insulated and uninsulated walls with the same U values. The table showed the variation of thermal performance depending on the location of the insulating membrane, the best result being with it positioned on the inner surface, as the surface temperature and mean radiant temperature build up quickly, although its effect on energy use is most marked with intermittent heating.

This is in conflict with the house described in a previous chapter by Mr Wilson where insulation was placed on the outside, thus using the thermal capacity of the structure. The importance of assessing accurately the energy input requirements for heating buildings is recognised. A computer method of investigating thermal behaviour of buildings has been developed by British Gas known as THERM which is being made generally available. This is not merely a computerised method of carrying out the conventional heat loss calculations, but a completely new approach taking into account transient conditions, with allowance for varying weather and heat gains from occupancy, lighting, machinery, etc.

Professor P. Burberry
written comment

The problem of insulation position is one which is very important but as yet unresolved. Internal insulation gives a fast response and if the heating installation can also respond quickly there are good prospects for economy in this respect. On the other hand external insulation gives additional internal thermal storage which can contribute to more equable temperatures and the better utilisation of solar and wild heat gains. It is probable that for winter economy internal insulation has advantages while for summer comfort external insulation may be better but the planning, orientation and fenestration of the building, the thermal properties of the materials and the resistance of materials to temperature variations and moisture contents are all critical factors. It is important that research should be brought to bear to establish the ideal solution of the problem.

I am surprised that Mr Peters says that the THERM computer program is a completely new approach. The factors he gives of varying weather and heat gains from occupancy were taken into account in the predictions which formed the basis of my tables and I am sure that many other people would say that they had also taken these factors into account.

Dr. N. S. Sturrock (Liverpool Polytechnic)
written comment

I am concerned that the myth that south-facing glazing is energy self-sufficient is being perpetuated. Conclusions drawn from Table 2 of Chapter 13 are suspect. Table 2 appears to be based on BRS Digest No 94 (First Series), 1956.[1] The U value of 3.4 $W/m^2 °C$ for a window, curtained at night, should be applied to the total window area, not just the glazed area of the window. Danter and Dick assumed that 30% of a domestic window was taken up by a wood frame and this was included in arriving at the value for the thermal transmittance quoted. The solar heat gain through such a window, per unit area of window (not glass), is only 70% of the amount given in Table 2.

The month which gives the greatest gain for a south-facing window is September, shown in Fig 8(c), and the question arises as to the extent to which the month of September can be considered to be part of the heating season. Danter and Dick assumed a heating season of 33 weeks in their paper, but dates were not specified. Even if mid-September is taken to be the commencement of the heating season, the gains indicated in Fig 8(c) are greatly reduced.

Professor Burberry has said elsewhere[2] that for all the solar energy that enters a room to be useful, any arriving at a time when the temperature of the room air is above the design temperature must be absorbed into the fabric and released only when required. This will not be the case in practice, even with a heavyweight building.[3]

[1] based on Danter and Dick paper to the Institute of Fuel Conference on Domestic Heating, May, 1956.
[2] Burberry P., *Architects Journal*, 4 Feb 1976, P 248.
[3] Davies M.G., Sturrock N.S., Benson A.C., 'Some results of measurements in St George's School, Wallasey. JIHVE, July 1971, p 84.

Professor P. Burberry
written comment

Dr Sturrock makes three points. One is about the complexity of arriving at a specific U value for a window curtained at night. Not only would this be subject to the area of the wooden frame, it would also be subject to the time of drawing the curtains, the weight of the curtain material and a number of other factors. However, the variation which is likely is probably comparatively small in relation to the variability of the weather and the pattern of internal use of different buildings. It seems reasonable to use the value suggested by BRE as a practical design figure.

Dr Sturrock, very correctly, draws attention to the fact that heat gained through a window is not all useful. In the spring and autumn, and even in the depth of the winter, the amount of heat coming through a window may well cause overheating which, if the heating installation cannot be turned off, can only be controlled by opening windows. If maximum advantage is to be taken of solar heat gain through windows it is important that the heating installation is readily amenable to control, preferably automatic, and that the fabric and finishes of the building are capable of absorbing and storing the heat entering at peak times and retransmitting it into the room later when it is needed. Astonishingly, little study has gone into this aspect of storing and utilising solar heat which is very surprising considering that the cost benefit from such studies would be extremely high.

The conclusion to be drawn from Table 2 is not that windows should be made large on southern orientations but that an effort should be made to plan and orient the building so that windows of sizes which are required are located on the best possible orientation for heat balance. The table makes the importance of this very clear. It is astonishing to see the way in which windows are, in very many cases, oriented at random without regard to the thermal consequences. This is an aspect of design where some improvement of the thermal performance of buildings could be achieved without any cost whatever.

Summary

G. Grenfell-Baines

Mr Dick's helpful advice in Chapter 1 to keep options open implies anticipatory design. It seems that the use of ambient energy may well be through conversion to electricity and this should be taken into account in designing for the years ahead. Many fossil fuel appliances being installed now have a life of only about fifteen years and in an evolutionary situation we need to think of the next phase. Individual contributions count in the evolutionary process. For example, individual attempts to apply total energy appear to be successful even though the concept has not yet been used in a nationally co-ordinated way. Such individual attempts, and there are many in the field of ambient energy, can be likened to the drops of water that together make up an ocean.

The strategy diagram (see page 76), with its implications that standards should be maintained while making savings out of waste, is a valuable contribution and something which could be shown to clients.

Perhaps the most important implication of the papers and discussion is the need for collaboration. Energy management is a problem with many facets and each requires specialist knowledge. To draw specialist knowledge together is a difficult task but can provide very good results when carried through successfully. Architects must realise the value of collaboration and be effective generalists, not stylists. Specialism in various fields has taken us far, but combinations of specialisms are required.

The flexible approach to economics is important, particularly the approach to uncertainty and the ability to take probability into account. It is hardly reasonable to expect a return on expenditure in five years, mortgage repayment periods might be a more appropriate guide.

One sympathises with the need to provide data on solar collector performance expressed by several contributors. In an area of uncertainty and experiment it might well be best to simply set limits, to define ranges.

Storage is obviously an important element in dealing with the vagaries of ambient energy. On the one hand extra volume is required for storage equipment whilst on the other we are urged to reduce the volume of buildings. More study is needed in this area and one wonders if the fabric of a house could be enlisted as storage. Floors are currently used as panels for warming and similar arrangements might be used as thermal stores.

The integration of the various contributions into total design has been raised and a sentiment I liked was "find out what has to be done and do it delightfully". We have to discover how the elements can give expression to form, how they can excite surfaces and even silhouettes. The visual aspect matters and will, I believe, inspire design as we learn more. The wind wall described in Chapter 8, with its sloping approaches incorporating solar panels, represents an integration towards total design.

Critical design decisions will be those which influence the balance between centralisation and decentralisation. Methods of collection, distribution and control must be evaluated and a balance reached between efficiency and humanity. Humanity is very much a matter of scale, but scale is not simply size. After reading Schumacher's "Small is Beautiful"* I felt I could write a book called "And Big Can Be Beautiful Too". It is a matter of proportion when deciding priorities and a sense of proportion is needed in reaching overall solutions.

Whether to use 'high technology', 'low technology' or 'alternative technology' are not the real questions. Technology is an instrument, it is our objectives and our approach to achieving them which are the vital subjects for discussion. Effective energy management is perhaps the most vital objective and "learning to live with nature" another. Exactly what technology is employed depends upon the objectives.

*Schumacher, E F, "Small is Beautiful – a Study of Economics as if People Mattered", Blond and Briggs, London, 1973.

Index

Where a topic is mentioned several times under the same heading, only the first occurrence has normally been cited.

0